ENERGY, RESOURCES, and the LONG-TERM FUTURE

World Scientific Series on Energy and Resource Economics

(ISSN: 1793-4184)

Published

Vol. 1 Quantitative and Empirical Analysis of Energy Markets
by Apostolos Serletis

Vol. 2 The Political Economy of World Energy: An Introductory Textbook
by Ferdinand E Banks

Vol. 4 Energy, Resources, and the Long-Term Future
by John Scales Avery

Forthcoming

Vol. 3 Bridges Over Water: Understanding Transboundary Water Conflict, Negotiation and Cooperation
by Ariel Dinar, Shlomi Dinar, Stephen McCaffrey & Daene McKinney

World Scientific Series on
Energy and Resource Economics – Vol. 4

ENERGY, RESOURCES, and the LONG-TERM FUTURE

John Scales Avery

H.C. Ørsted Institute,
University of Copenhagen, Denmark

NEW JERSEY • LONDON • SINGAPORE • BEIJING • SHANGHAI • HONG KONG • TAIPEI • CHENNAI

Published by

World Scientific Publishing Co. Pte. Ltd.
5 Toh Tuck Link, Singapore 596224
USA office: 27 Warren Street, Suite 401-402, Hackensack, NJ 07601
UK office: 57 Shelton Street, Covent Garden, London WC2H 9HE

British Library Cataloguing-in-Publication Data
A catalogue record for this book is available from the British Library.

ENERGY, RESOURCES, AND THE LONG-TERM FUTURE
World Scientific Series on Energy and Resource Economics — Vol. 4

ISBN-13 978-981-270-764-2
ISBN-10 981-270-764-6

Printed in Singapore.

Contents

Preface

This book is dedicated to Dr. Jens Junghans, who for many years has been concerned with the problem of industrial society's totally unsustainable relationship with nature. I am fortunate to have had frequent opportunities to discuss this issue with Dr. Junghans, and I am convinced of it's importance. With his permission, I am using some of his ideas in this book, and adding a few of my own.

To a large extent, this book is a history of the relationship between humankind and the natural world, and to a lesser extent it attempts to predict the long-term consequences of our flawed relationship with nature. In other words, a study of human folly in the past and present is used as a springboard for making predictions about the future problems.

It is notoriously difficult to make correct predictions about the distant future. In modern human society, the breakneck speed of scientific discovery and technological innovation makes long-term predictions especially difficult. Furthermore, people tend to care only about those things that will occur only during their own lifetimes. For these reasons, among others, most economists do not look more than a few decades into the future.

Nevertheless, I believe that the distant future of the balance or unbalance between humankind and nature has a great importance. Certainly, if we look far enough ahead, it will be beyond our own lifetimes, but we have a great responsibility to our descendants.

Looking at the very distant future simplifies some issues. For example, one can argue about the size of reserves of coal, oil and metals, but it is certain that in the very long run, such non-renewable resources will become rare and extremely expensive.

Viewed on a time scale of many thousands of years rather than tens or hundreds, global population growth and fossil fuel use appear in an extremely clear and dramatic perspective. Forty thousand years ago, at the time when human cultural development began to accelerate, there were at most only 4-5 million or so members of our species on the earth. They lived as hunter-gatherers, and were not conspicuously different from other animals. Then, suddenly, a series of cultural achievements allowed humans to increase enormously in numbers and to populate all parts of the earth. The invention of agriculture was followed by the inventions of writing, paper, and printing. Knowledge, giving humans mastery over the natural world, began to accumulate with astonishing rapidity. New advances in technique allowed further growth in population.

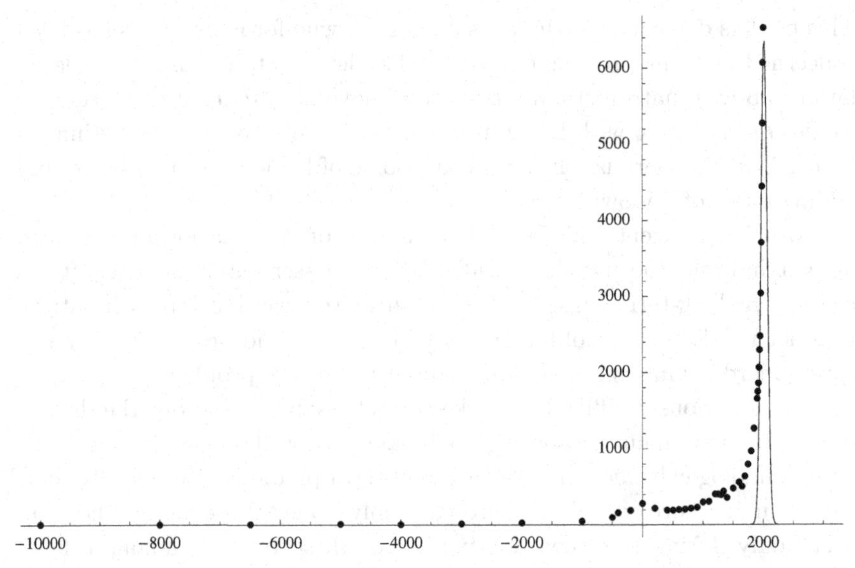

This figure shows the estimated human population as a function of time during the last 12,000 years. The dots are the US Census Bureau's population estimates in millions, while the solid curve shows the approximate time-dependence of fossil fuel use. When they are plotted together, the sudden, explosive growth of human population, and the spike-like graph of fossil fuel use, are seen to be simultaneous (and causally connected).

Plotted on an evolutionary timescale, human population growth appears as an extraordinarily abrupt upward surge. On the same time scale, a graph of fossil fuel use is a tall, narrow spike, rising from zero to a high value, and then falling to nothing again, all in the space of a few centuries. Looking at the figure above, we can infer that fossil fuel use has been one of the causes of the explosive upsurge of global population.

One can calculate from the size of coal, oil and natural gas reserves that the era of fossil fuel use will end within a few hundred years. Must the graph of human population also fall abruptly at the same time? This is one of the questions that we will address.

The Industrial Revolution (when both fossil fuel use and explosive population growth began in earnest) was the result of the rapid accumulation of human knowledge following the invention of printing with movable type. In Europe, the rise of science in the 17th and 18th centuries produced a period of great optimism. The Utopian predictions of the Marquis de Condorcet, William Godwin and Adam Smith (Chapter 1) are typical of this period. In the early 19th century, however, the realities of persistent poverty led Thomas Robert Malthus and David Ricardo (Chapter 2) to a much darker picture of the human condition - so dark, in fact, that it led Thomas Carlyle to call economics "the Dismal Science".

Since the time of Adam Smith, industrial society has thundered forward under the banner of unrestricted economic growth that Smith was the first to raise. Today, however, as we approach limits to growth imposed by the exhaustion of non-renewable resources and by the finite carrying capacity of the global environment, we should perhaps listen also to the warning voice of Malthus. He pointed out that throughout almost all of human history, the growth of population has been held in check by strong forces. These are sometimes preventative checks, such as late marriage, moral restraint or contraception (which he called "vice"); but when the preventative checks fail, the grim Malthusian forces - famine, disease and war - come into play.

The successes of science and technology have allowed dramatic growth of both population and economic activity during the last few centuries, but the limits to both types of growth are rapidly approaching. It is therefore relevant to ask what level of global population and what level of economic activity can be comfortably sustained in the distant future.

A stable future world must necessarily be a war-free world, since weapons are likely to become even more destructive in the future than they are today. A world war fought with such weapons would destroy civilization. Thus our descendants will also be faced with the great task of abolishing the institution of war. They will not only need to stabilize and eventually reduce global population and economic activity; they will also need to develop political and ethical maturity to match their scientific progress.

In arguing against the optimistic Utopia of William Godwin, Malthus says, "The great error under which Mr. Godwin labours throughout his whole work is the attributing of almost all of the vices and misery that prevail in civil society to human institutions. Political regulations and established administration of property are, with him, the fruitful sources of all evil, the hotbeds of all the crimes that degrade mankind. Were this really the true state of the case, it would not seem a completely hopeless task to remove evil completely from the world; and reason seems to be the proper and adequate instrument for effecting so great a purpose. But the truth is, that although human institutions appear to be, and often are, the obvious and obtrusive causes of much misery in human society, they are, in reality, light and superficial in comparison with those deeper-seated causes of evil which result from the laws of nature and the passions of mankind."

In the passage just quoted, Malthus is talking about population growth: The passions of mankind drive us to excessive reproduction, while the laws of nature limit our food supply. Hence the poverty and misery that we observe in many parts of the world. However, his words could equally well be applied to the problem of war: The laws of nature make nuclear weapons possible, and the passions of mankind drive us to use them.

Thus we are faced with the question of whether human reason can lead us to a stable and happy world in the distant future, or whether the laws of nature and the passions of mankind will make impossible all of the future Utopias that one can visualize.

At the time when Malthus was writing, Darwin's theory of evolution had not yet appeared, but today, with our knowledge of evolutionary theory, we must ask how "the passions of mankind" (i.e. human emotional nature) can be so counterproductive. At first sight, one would think that evolutionary forces would make humans well-adapted their environment and way of life in much the same way that swallows are adapted to catching gnats or storks adapted to a diet of frogs. The answer to this seeming paradox lies in the extreme slowness of genetic evolution compared with the rapidity of human cultural evolution.

Before cultural evolution began to revolutionize the lifestyle of our species, the "passions of mankind" were undoubtedly necessary for the survival of our remote ancestors. However, the rapid and constantly accelerating rate of cultural evolution has changed the conditions of human life beyond recognition during the last forty thousand years. Genetically we are very similar to our hunter-gatherer ancestors, but their world has been replaced by a world of quantum theory, space travel, gene splicing and information technology. Thus human emotions, which have remained relatively unchanged, are often inappropriate for our present way of life. In the future, the problem of anachronistic emotions is certain to become even more acute. This book will discuss the educational measures that will be needed to overwrite the destructive passions of mankind.

If we carefully examine cultural evolution, we can see that it has two parts, one of which changes more quickly than the other. The extremely rapidly-moving part is science and technology. Our political and social institutions change more slowly, although their progress is still very rapid compared with genetic change. Because of the different rates of change of these two facets of cultural evolution, our political and social institutions often fail to harmonize with the innovations of science and technology. For example, in a world of thermonuclear weapons, the absolutely sovereign nation-state has become a dangerous anachronism - yet it persists because of institutional inertia. It takes quite a bit of time for laws, constitutions, schoolbooks, thought-patterns and political structures to adjust to new realities. In the meantime, technology roars ahead, with a rate of change so great that it threatens to shake society to pieces.

Thus modern human society experiences two types of tensions, both of which will probably become more acute in the future:

(1) Tensions produced by the fact that our emotions do not harmonize with our present way of life.
(2) Tensions produced by the disharmony between our technology and our social and political institutions.

How can we find the path to a stable and peaceful future society? Can we take some steps along this path today? Can industrial society achieve equilibrium and harmony with nature, as well as harmony among all the humans that inhabit the earth? Let us explore these questions together.

Acknowledgments

I would like to begin by thanking Dr. Jens Junghans for the original inspiration, and for his continued advice and help during the writing of this book. This help included the loan of many books from his large personal library. Next I want to express my gratitude to my son James for the enormously valuable technical advice that he has given to me. I am also extremely grateful to Dr. Klaus Illum, Mrs. Ruth Gunnarsen, Mr. Jan Møller, Mr. Mark Kamio, Dr. Tom Børsen Hansen, Dr. Jens-Christian Navarro Poulsen, and economist Michael Foote for their helpful comments.

Chapter 1

THE IDEA OF PROGRESS

1.1 Cultural evolution

When our ancestors began to evolve a complex language and culture, it marked the start of an entirely new phase in the evolution of life on earth. In all terrestrial organisms, information is transmitted between between generations by means of the genetic code; and genetic evolution takes place through natural selection acting on modifications of this code. In human cultural evolution, information is also transmitted between generations by means of language. This second mode of evolution gave our species enormous adaptive advantages. While genetic changes are random and slow, cultural changes are purposeful and rapid. For example, when our ancestors moved out of Africa and spread over Europe and Asia, they did not adapt to the colder climate by growing long fur, but instead invented clothing.

An acceleration of human cultural development seems to have begun approximately 40,000 years ago. The first art objects date from that period, as do migrations that ultimately took modern man across the Bering Strait to the western hemisphere. A land bridge extending from Siberia to Alaska is thought to have been formed approximately 70,000 years ago, disappearing again roughly 10,000 years before the present. Cultural and genetic studies indicate that migrations from Asia to North America took place during this period.

The agricultural revolution

Beginning about 10,000 B.C., the way of life of the hunters was swept aside by a great cultural revolution: the invention of agriculture. The earth had entered a period of unusual climatic stability, and this may have helped to make agriculture possible. The first agricultural villages date from this time, as well as the earliest examples of pottery. Dogs and reindeer were domesticated, and later, sheep and goats.

Radio-carbon dating shows that by 8500 B.C., people living in the caves of Shanidar in the foothills of the Zagros mountains in Iran had domesticated sheep. By 7000 B.C., the village farming community at Jarmo in Iraq had domesticated goats, together with barley and two different kinds of wheat.

Starting about 8000 B.C., rice came under cultivation in East Asia. This may represent an independent invention of agriculture, and agriculture may also have been invented independently in the western hemisphere, made possible by the earth's unusually stable climate during this period.

At Jericho, in the Dead Sea valley, excavations have revealed a prepottery neolithic settlement surrounded by an impressive stone wall, six feet wide and twelve feet high. Radiocarbon dating shows that the defenses of the town were built about 7000 B.C.. Probably they represent the attempts of a settled agricultural people to defend themselves from the plundering raids of less advanced nomadic tribes.

Starting in western Asia, the neolithic agricultural revolution swept westward into Europe, and eastward into the regions that are now Iran and India. By 4300 B.C., the agricultural revolution had spread southwest to the Nile valley, where excavations along the shore of Lake Fayum have revealed the remains of grain bins and silos. The Nile carried farming and stock-breeding techniques slowly southward, and wherever they arrived, they swept away the hunting and food-gathering cultures. By 3200 B.C. the agricultural revolution had reached the Hyrax Hill site in Kenya. At this point the southward movement of agriculture was stopped by the swamps at the headwaters of the Nile. Meanwhile, the Mediterranean Sea and the Danube carried the revolution westward into Europe. Between 4500 and 2000 B.C. it spread across Europe as far as the British Isles and Scandinavia.

Early forms of writing

The speed of human cultural evolution leapt to an entirely new level with the invention of writing. Suddenly, with this critically important invention, information began to accumulate at an unprecedented rate.

In Mesopotamia (which in Greek means "between the rivers"), the settled agricultural people of the Tigris and Euphrates valleys evolved a form of writing. The practical Mesopotamians seem to have invented writing as a means of keeping accounts.

Small clay and pebble counting tokens symbolizing items of trade began to be used in the Middle East about 9000 B.C., and they were widely used in the region until 1500 B.C.. These tokens had various shapes, depending on the ware which they symbolized, and when made of clay they were often marked with parallel lines or crosses, which made their meaning more precise. In all, about 500 types of tokens have been found at various sites. Their use extended as far to the west as Khartoum in present-day Sudan, and as far to the east as the region which is now Pakistan. Often the tokens were kept in clay containers which were marked to indicate their contents. The markings on the containers, and the tokens themselves, evolved into true writing.

Among the earliest Mesopotamian writings are a set of clay tablets found at Tepe Yahya in southern Iran, the site of an ancient Elamite trading community halfway between Mesopotamia and India. The Elamite trade supplied the Sumerian civilization of Mesopotamia with silver, copper, tin, lead, precious gems, horses, timber, obsidian, alabaster and soapstone.

The tablets found at Tepe Yahya are inscribed in proto-Elamite, and radiocarbon dating of organic remains associated with the tablets shows them to be from about 3600 B.C.. The inscriptions on these tablets were made by pressing the blunt and sharp ends of a stylus into soft clay. Similar tablets have been found at the Sumerian city of Susa at the head of the Tigris River.

In about 3100 B.C. the cuneiform script was developed, and later Mesopotamian tablets are written in cuneiform, a phonetic script in which the symbols stand for syllables.

The Egyptian hieroglyphic (priest writing) system began its development in about 4000 B.C.. At that time, it was pictorial rather than phonetic. However, the Egyptians were in contact with the Sumerian civilization of Mesopotamia, and when the Sumerians developed a phonetic system of writing in about 3100 B.C., the Egyptians were quick to adopt the idea. In the cuneiform writing of the Sumerians, a character stood for a syllable. In the Egyptian adaptation of this idea, some of the symbols stood for syllables or numbers, or were determinative symbols, which helped to make the meaning of a word more precise. However, some of the hieroglyphs were purely alphabetic, i.e. they stood for sounds which we would now represent by a single letter. This was important from the standpoint of cultural history, since it suggested to the Phoenicians the idea of an alphabet of the modern type.

In Sumer, the pictorial quality of the symbols was lost at a very early stage, so that in the cuneiform script the symbols are completely abstract. By contrast, the Egyptian system of writing was designed to decorate monuments and to be impressive even to an illiterate viewer; and this purpose was best served by retaining the elaborate pictographic form of the symbols.

Starting with the neolithic agricultural revolution and the invention of writing, human culture began to develop with explosive speed. Agriculture

Fig. 1.1 *A baked clay tablet with cuneiform writing.*

led to a settled way of life, with leisure for manufacturing complex artifacts, and for invention and experimentation. Writing allowed the cultural achievements of individuals or small groups to become widespread, and to be passed efficiently from one generation to the next.

Compared with the rate of ordinary genetic evolution, the speed with which the information-driven cultural evolution of Homo sapiens sapiens began to develop is truly astonishing. 12,000 years before the present, our ancestors were decorating the walls of their caves with drawings of mammoths. Only 10,000 years later, they were speculating about the existence of atoms! Within another 2000 years, they could see individual atoms in scanning tunneling microscopes. New methods for the conservation, transmission and utilization of information were the driving forces behind the explosively accelerating evolution of human culture.

This remarkably rapid growth of human culture was not accompanied by very great genetic changes in our species. It took place instead because of a revolutionary leap in the efficiency with which information could be conserved and transmitted between generations, not in the code of DNA, but in the codes of Mesopotamian cuneiform, Egyptian hieroglyphics, Chinese ideograms, Mayan glyphs, and the Phoenician and Greek alphabets.

The development of printing in Europe and the rapid spread of books and knowledge produced a brilliant chainlike series of scientific discoveries - the sun-centered system of Copernicus, Kepler's three laws of planetary motion, Descartes' invention of analytic geometry, Gilbert's studies of magnetism, Galileo's discoveries in experimental physics and astronomy, the microscopy of Hooke and Leeuwenhoek, Newton's universal laws of motion and gravitation, the differential and integral calculus of Newton and Leibniz, the medical discoveries of Harvey, Jenner, Pasteur, Koch, Semmelweis and Lister, and the chemical discoveries of Boyle, Dalton, Priestly, Lavoisier and Berzelius.

The rapid accumulation of scientific knowledge made possible by paper and printing was quickly converted into the practical inventions of the industrial revolution. In the space of a few centuries, the information explosion changed Europe from a backward region into a society of an entirely new type, driven by scientific and technological innovation and by the diffusion and accumulation of knowledge.

1.2 Condorcet

The rise of science in Europe which followed the invention of printing with movable type inspired a period of optimism - the Enlightenment. In the 18th century, applications of scientific knowledge were already being applied to practical problems, with the result that great improvements in the human condition seemed within reach. The Utopian writings of the marquis de Condorcet, William Godwin and Adam Smith are typical of the period.

Marie-Jean-Antoine-Nicolas Caritat, marquis de Condorcet, was one of the first people to write about evolution, and his ideas embraced both its forms - both genetic evolution and cultural evolution. He born in 1743 in the town of Ribemont in southern France. He studied mathematics in Paris, and in 1765, when he was barely 22 years old, he presented an *Essay on the Integral Calculus* to the Academy of Sciences in Paris.

Condorcet was interested in social questions as well as mathematics, and he became one of the chief authors of the proclamation which declared France to be a republic and which summoned a National Convention. He also was one of the authors of a draft of a new French Constitution. This brought him into conflict with the Jacobin leader Robespierre, who had helped to write a competing constitution.

When Condorcet urged Frenchmen to reject the Jacobin version of the constitution, Robespierre succeeded in having him denounced in the Convention; and an order was sent out for his arrest. The officers tried to find

Fig. 1.2 *A printing press ca. 1520.*

him, first at his town house, and then at his house in the country; but, warned by a friend, Condorcet had gone into hiding.

Although Robespierre's agents had been unable to arrest him, Condorcet was sentenced to the guillotine *in absentia*. He knew that in all probability he had only a few weeks or months to live; and he began to write his last thoughts, racing against time. Condorcet returned to a project which he had begun in 1772, a history of the progress of human thought, stretching from the remote past to the distant future. Guessing that he

Fig. 1.3 *Marquis de Condorcet (1743-1794).*

would not have time to complete the full-scale work he had once planned, he began a sketch or outline: *Esquisse d'un tableau historique des progrès de l'esprit humain.*

Like many other philosophers of the Enlightenment, Condorcet believed in the infinite perfectibility of humankind, and he anticipated some of Charles Darwin's evolutionary ideas: Condorcet believed that both humans and animals have evolved from lower forms and are still undergoing evolutionary improvement. He guessed that humans were once at the same level as animals are today; but in humans, higher facilities gradually developed. Since this improvement of our species took place historically, he believed, there is no reason why further evolution should not take place in the future.

Condorcet begins his manuscript by stating his belief "that nature has set no bounds on the improvement of human facilities; that the perfectibility of man is really indefinite; and that its progress is henceforth independent of any power to arrest it, and has no limit except the duration of the globe upon which nature has placed us". He states also that "the moral goodness of man is a necessary result of his organism; and it is, like all his other facilities, capable of indefinite improvement."

Isaac Newton's great work, *Philosophiae Naturalis Principia Mathamatica*, had been translated into French and was much discussed by the philosophers of the Enlightenment. Condorcet accepted Newton's concept of an orderly cosmos, governed by natural laws to which there are no exceptions, and he believed that these laws govern the existence and evolution of both humans and animals. In his *Esquisse*, he stresses the similarities between humans and animals, and their probable common origin. To understand how our species could have evolved from lower forms of life, Condorcet says, one need only imagine a series of small changes, continued over a long period of time.

The prolonged childhood of humans is unique among animals, Condorcet points out. He states that this prolonged childhood is necessary for the intergenerational transmission of speech and culture. As speech and culture evolved in humans, the need for a stable family structure also evolved, to protect the young during their long infancy and education. Thus, biological evolution produced a moral precept - the sanctity of the family.

Condorcet then gives his readers another instance where biological evolution has produced moral behavior patterns in humans: As cultural evolution continued, he says, large groups of humans needed to be able to live together, and this would have been impossible without some degree of sensitivity to the needs of others. Thus, Condorcet says, altruism in humans has its origin in biological and cultural evolution.

Unlike Rousseau, who saw civilized humans as degraded when compared to the "noble savage", Condorcet believed the development of civilized societies to be an upward step. "The stormy and painful passage", he wrote, "of a primitive society to a state of enlightenment and freedom is by no means a degeneration of the human species, but a necessary crisis in its march towards a future state of absolute perfection. The reader will see that not the increase but the decay of knowledge has produced the vices of polished peoples; and finally that, far from corrupting mankind, knowledge has sweetened their temper even when it could not correct their faults or alter their character."

Condorcet believed that in the future, ideas that now require genius to comprehend will become part of the general education of humankind, because the ideas will become systematized and simplified. He was perhaps thinking of his own contributions to calculus, which had been developed through the genius of Newton and Leibniz. The work of Condorcet helped to turn calculus into an everyday tool that could be used by all scientists and engineers. He also predicted the development of the social sciences, and here again, Condorcet was a pioneer. He was the first to apply statistics to economic and social problems.

Condorcet believed that vice is largely due to ignorance and error, and that it can be eliminated through education and social change. He lists the errors that he has noticed in the human institutions of his time: inequality between men and women, economic inequalities, religious bigotry, war, slavery, disease, division of humanity into mutually exclusive linguistic groups, and hereditary transmission of power.

Condorcet regards the republican form of government as being greatly superior to monarchy, and he looks forward to the time when democracies will be established throughout the world. This will reduce the probability of wars between nations, and will also end the sufferings experienced under tyrannical monarchs.

In a previous publication, *Sur L'Admission des Femmes au Droit de Cité*, (Paris, 1790), Condorcet had pleaded for political and educational equality between men and women, and he returns to this theme in his *Esquisse*. Condorcet sees no reason, moral, physical or intellectual, for inequality between the sexes, and he regard it as a defect in the French Constitution that women are not given full political rights.

Regarding disease, Condorcet predicts that the progress of medical science will abolish it. Also, he maintains that since perfectibility (i.e. evolution) operates throughout the biological world, there is no reason why mankind's physical structure might not gradually improve, with the result that human life in the remote future could be greatly prolonged.

Condorcet looks forward to a time when the atrocious suffering produced by war will disappear - a time when war will be abolished as a social institution. With their intellectual improvement, he says, humans will come to recognize war as an enormous source of unnecessary destruction and pain. The establishment of republican governments throughout the world will end dynastic conflicts. Wars fought for commercial reasons will also disappear. Finally, the introduction of a universal language throughout the world will lead to increased understanding between nations. Peace will be further strengthened by perpetual confederations.

Condorcet looks forward to a time when both colonial exploitation and the slave trade will be abolished. He also predicts that sometime in the future, all unjust monopolies (like the Dutch and English East India Companies) will be prohibited, and the benefits of free trade will be extended throughout the world. He advocates a system of universal, state-supported education, as well as insurance schemes to provide pensions for widows and orphans. The result of these social reforms, he says, will be to minimize social and financial inequalities between classes.

Realizing that as the conditions of life become more favorable, population will increase, Condorcet writes, "...of industry and happiness, each generation will be called to more extended enjoyments, and in consequence, by the physical constitution of human frame, to an increase in the number of individuals. Must not a period then arrive when these laws, equally necessary, shall contradict each other; when the increase in the number of men surpassing their means of subsistence must necessarily result in either a continual diminution of happiness and population - a movement truly retrograde; or at least a kind of oscillation between good and evil? In societies arrived at this term, will not this oscillation be a constantly subsisting cause of periodical misery? Will it not mark the limit when all further melioration will become impossible, and point out that boundary to the perfectibility of the human race, which it may reach in the course of ages, but can never pass?"

"There is no person", he continues, "who does not see how very distant such a period is from us. But shall we ever arrive at it? It is equally impossible to pronounce for or against the future realization of an event which cannot take place but in an era when the human race will have attained improvements, of which we can at present scarcely form a conception."

When the time finally arrives when global overpopulation threatens to limit the possibility of human progress, Condorcet says, mankind will have reached such a high level of enlightenment that superstitions will have vanished. At that time (which Condorcet imagines to be in the distant future), humans will be so enlightened that they will find a solution to the problem of overpopulation, either through contraception or promiscuity. It may seem strange that he proposes promiscuity as a means for reducing the birth rate, but in fact, minor venereal diseases can lead to sterility if they are untreated, as they were at the time when Condorcet was writing.

With his own life in mortal danger, Condorcet nevertheless ends his *Esquisse* on a serene note: He says that a person who has contributed to human progress to the best of his or her ability has nothing to fear from fate. We can take comfort in the vision of humanity's inevitable march towards a better future.

Shortly after Condorcet completed the *Esquisse*, he was arrested and died in prison. After Condorcet's death the currents of revolutionary politics shifted direction. Robespierre, the leader of the Terror, was himself soon arrested. The execution of Robespierre took place on July 25, 1794, only a few month's after the death of Condorcet.

Condorcet's *Esquisse d'un tableau historique des progrès de l'esprit humain* (*Sketch of an Historical Picture of the Progress of the Human Spirit*) was published in 1795, a year after the death of its author. The Convention voted funds to have it printed in a large edition and distributed throughout France, thus adopting the *Esquisse* as its official manifesto. Today the concepts put forward in Condorcet's *Esquisse* still influence our understanding of human progress. As we shall see, the small but influential book also provoked Malthus to write *An Essay on the Principle of Population.*

1.3 Godwin

William Godwin's book *Political Justice* was published in England in 1793. Like Condorcet's *Esquisse*, it not only helped to define the idea of progress but also helped to provoke Malthus' essay on population.

William Godwin was an English political writer and novelist who was much influenced by the French philosophers of the Enlightenment. In *Political Justice*, he maintains that the time has come for a thorough reform of society. Every social institution without exception must be examined in the light of reason, and must be discarded if found to be corrupt, irrational or unjust. No institution is safe or sacred just because it is ancient.

In *Political Justice*, Godwin insists that politics, ethics and knowledge are closely linked. His book is an enthusiastic vision of what humans could be like at some future period when the trend towards moral and intellectual improvement has lifted men and women above their their present state of ignorance and vice. Much of the harshness of the penal system will then be unnecessary, Godwin believes. (At that time there were more than a hundred capital offenses in England, and this number had soon increased to almost two hundred. The theft of any object costing more than ten shillings was punished by hanging.)

Present society, Godwin says, holds most of its citizens "in abject penury, rendered stupid with ignorance and disgustful with vice, perpetuated in nakedness and hunger, goaded to the commission of crimes, and made victims to the merciless laws which the rich have instituted to oppress them". But human behavior is subject to natural laws no less than the planets of Newton's solar system. "In the life of every human", Godwin writes, "there is a chain of causes, generated in that eternity which preceded his birth, and going on in regular procession through the whole period of his existence, in consequence of which it was impossible for him to act in any instance otherwise than he has acted." If the conditions of upbringing, education and environment are improved, behavior will also improve.

Crime and vice should be treated in the same way that we treat disease. We should try to remove the causes of poverty, ignorance, vice and crime and attempt to cure human failings instead of punishing them. Thus, Godwin says, "our disapprobation of vice will be of the same nature as our disapprobation of an infectious distemper."

In defining the higher moral level towards which improved education and environment should aim, Godwin draws heavily on his Christian background, especially the Parable of the Good Samaritan: We must try to love our neighbor, even though he may not be a member of our family, circle of friends or nation. Although separated from us by cultural differences, ethnic background or geographical distance, he is still our neighbor, a member of the human family; and our duty to him is no less than our duty to those who are closest to us. It follows that narrow loyalties must be replaced or supplemented by loyalty to the interests of humanity as a whole.

In every act, Godwin maintains, we have a duty to weigh the amount of good which we are doing for the benefit of humanity. "I am bound", he writes, "to employ my talents, my understanding, my strength and my time for the production of the greatest quantity of the general good".

It is not the responsibility of the state to judge the benevolence of our actions. It is the responsibility of each individual conscience, Godwin says. The individual must follow his or her conscience even if it conflicts with the dictates of the state. We should struggle to change our institutions and laws if they fail to meet the criteria of benevolence, justice and truth. Anticipating Thoreau, Tolstoy and Gandhi, Godwin gives personal judgment a completely central role.

The duty of benevolence means that we only hold private property in trust. It should be used where it will do the most good: "To whom does any article, suppose a loaf of bread, justly belong?", Godwin asks, "... I have an hundred loaves in my possession, and in the next street there is a poor man expiring with hunger, to whom one of these loaves would be a means of preserving his life. If I withhold this loaf from him, am I not unjust? If I impart it, am I not complying with what justice demands? To whom does the loaf justly belong?"

Driven by the logic of his arguments, Godwin even goes so far as to deny that loyalty, gratitude and keeping promises are really virtues! He considers them to be vices because they can distort our judgment in evaluating the benevolence of our behavior. Loyalty, gratitude and promises might cause us to commit harmful acts that our individual consciences might otherwise forbid. An example of this is the suspension of private judgment that follows a soldier's induction into an army. (When the famous Whig parliamentarian Edmund Burke heard that Godwin considered gratitude to be a vice, he exclaimed, "I would save him from that vice by not doing him any service!")

Fig. 1.4 *William Godwin (1755-1836).*

At some point in the future, Godwin says, agriculture will be mechanized to such an extent that people will need only a few hours a day to earn their daily bread. Furthermore, war will be abolished, and luxuries will seem less important to the enlightened citizens of a future society. Since the struggle for existence, the struggle for power, and the struggle for social status will no longer be all-important, there will be room for generosity, culture and education.

Godwin describes his optimistic vision of a future society in the following words: "The spirit of oppression, the spirit of servility and the spirit of fraud - these are the immediate growth of the established administration of property. They are alike hostile to intellectual improvement. The other vices of envy, malice, and revenge are their inseparable companions. In a state of society where men lived in the midst of plenty, and where all shared alike the bounties of nature, these sentiments would inevitably expire. The narrow principle of selfishness would vanish. No man being obliged to guard his little store, or provide with anxiety and pain for his restless wants, each would lose his own individual existence in the thought of the general good. No man would be the enemy of his neighbor, for they would have nothing to contend; and of consequence philanthropy would resume the empire which reason assigns her. Mind would be delivered from her perpetual anxiety about corporal support, and free to expatiate in the field of thought which is congenial to her. Each man would assist the inquiries of all."

When *Political Justice* was published in 1793, France had just declared war on England, and England was preparing to make a counter-declaration against France. William Pitt the Younger, who was then Prime Minister, considered prosecuting both the author and the publisher, but reflecting that the book was 895 pages in length and cost more than three times the weekly wages of a laborer, he changed his mind, remarking that "a three-guinea book could never do much harm among those who have not three shillings to spare". However, Pitt was wrong in thinking that the price of Godwin's book would keep it from being widely read. In 1799 John Fenwick (Godwin's first biographer) wrote: "Perhaps no work of equal bulk ever had such a number of readers; and certainly no book of such profound inquiry ever made so many proselytes in an equal space of time".

Political Justice became best-seller, and pirated editions appeared in Ireland, Scotland, and America. Hundreds of groups of workers bought joint copies of the book, which were read aloud to meetings or circulated among subscribers. England was full of excited talk about the "New Philosophy" advocated in *Political Justice.*

Godwin achieved fame almost overnight. "I was nowhere a stranger", he wrote later, "...I was everywhere received with curiosity and kindness. If temporary fame ever was an object worthy to be coveted by the human mind, I certainly obtained it in a degree that has seldom been exceeded". As Godwin's friend, the essayist William Hazlitt, described it, "...he blazed as a sun in the firmament of reputation; no-one was more talked of, more looked up to, more sought after, and wherever liberty, truth, justice was the theme, his name was not far off." A few years later, the atmosphere of optimism in

England changed to disappointment and reaction, and Godwin's temporary fame evaporated.

1.4 Adam Smith

The optimism of the Enlightenment, which we have seen in the writings of Condorcet and Godwin, can also be found in the books of the great Scottish economist Adam Smith (c.a 1723-1790). Smith was baptized in 1723 in the town of Kirkcaldy, County Fife. His father, who was the town's controller of customs, had died six months previously. At the age of 4, the boy was kidnapped by a band of Gypsies, but he was rescued by an uncle and returned to his mother. The rescue was fortunate since, as his biographer later remarked, Adam Smith "would have made a poor Gypsy".

In 1737 (at the age of 14) Smith began the study of moral philosophy at Glasgow University, and in 1740 he obtained a scholarship to Balliol College, Oxford. He left Oxford in 1746 and two years later he began delivering public lectures in Edinburgh. Some of these lectures were devoted to "the progress of opulence" under "the obvious and simple system of natural liberty". Much later in his life he returned to this theme, which he developed in his most famous book, *Inquiry into the Nature and Causes of the Wealth of Nations.*

In 1751, Adam Smith was appointed professor of logic at the University of Glasgow, and in 1752 he exchanged this title for a chair in moral philosophy. At the university, Smith became locally famous not only for his wide-ranging scholarship, but also for his absent-mindedness and eccentricity. Although he was shy and hesitating at the start of his lectures, Smith usually warmed to the subject and became both eloquent and profound. Small busts of him were soon on sale at bookseller's shops. He became the close friend of the philosopher David Hume. Joseph Black and James Hutton, two of the most important scientists of the Scottish Enlightenment, were also among Smith's close friends.

In 1759, Adam Smith published a book entitled *The Theory of Moral Sentiments*, which was subtitled: *An Essay towards an Analysis of the Principles by which Men naturally judge concerning the Conduct and Character, first of their Neighbors, and afterwords of themselves.* In this book, he pointed out that people can easily judge the conduct of their neighbors. They certainly know when their neighbors are treating them well, or badly. Having learned to judge their neighbors, they can, by analogy, judge their own conduct. They can tell when they are mistreating their neighbor or being kind by asking themselves: "Would I want him to do this to me?" As Adam Smith put it:

"Our continual observations upon the conduct of others insensibly lead us to form to ourselves certain general rules concerning what is fit and proper to be done or avoided... It is thus the general rules of morality are formed."

When we are kind to our neighbors, they maintain friendly relations with us; and to secure the benefits of their friendship, we are anxious to behave well towards other people. Thus, according to Adam Smith, enlightened self-interest leads men and women to moral behavior.

The Theory of Moral Sentiments was a great success, and the book established an international reputation for its author. It also attracted the admiration of Charles Townshend (who is famous for precipitating the American Revolution through his heavy-handed policies as Chancellor of the Exchequer). Townshend had married the wealthy widow of the Duke of Buccleuch, and he was looking for a tutor for the stepson that he had acquired in this marriage. After being introduced to Adam Smith by David Hume, Townshend decided that Professor Smith would be a perfect tutor for the young Duke. He offered Smith a salary of 300 pounds per year (in those days a very considerable sum) as well as the prospect of an annual pension of 300 pounds when the tutoring duties were finished.

This was too good an offer to refuse, and in 1764 Smith resigned his position at Glasgow University and accompanied the young Duke on the Grand Tour of Europe, which was in those days considered to be an important part of the education of a nobleman. The tour gave Adam Smith a chance to meet François Quesnay, who, besides being a physician, was also the most famous French economist of the time, an opponent of Mercantilism and an advocate of economic freedom from governmental interference (*laissez-faire*).

In Quesnay's view, the mercantilist emphasis on bullion was wrong. Wealth, he believed, springs from production and flows through society in much the same way that blood flows through our arteries and nourishes the parts of the body. Smith agreed with this picture, but while Quesnay emphasized agriculture, Smith realized that industrial production is also an important source of wealth.

The tour was finally ended by a tragedy. The Duke's younger brother joined them, but he contracted a fever from which he died, despite Quesnay's frantic efforts to save him. After this sad event, Adam Smith and the young Duke returned to England. The pension provided by Townshend allowed Smith to retire to his mother's home in Kirkcaldy. Here, for the next ten years, he worked on the enormous book that he had begun in France.

In 1776, Adam Smith published his magnum opus, *Inquiry into the Nature and Causes of the Wealth of Nations*. In this book, he examined the reasons why some nations are more prosperous than others. Adam

Smith concluded that the two main factors in prosperity are division of labor and economic freedom.

As an example of the benefits of division of labor, he cited the example of a pin factory, where ten men, each a specialist in a particular manufacturing operation, could produce 48,000 pins per day. One man drew the wire, another straightened it, a third pointed the pins, a fourth put on the heads, and so on. If each man had worked separately, doing all the operations himself, the total output would be far less. The more complicated the manufacturing process (Smith maintained), the more it could be helped by division of labor. In the most complex civilizations, division of labor has the greatest utility.

Adam Smith believed that the second factor in economic prosperity is economic freedom, and in particular, freedom from mercantilist government regulations. He believed that natural economic forces tend to produce an optimum situation, in which each locality specializes in the economic operation for which it is best suited.

Smith believed that when each individual aims at his own personal prosperity, the result is the prosperity of the community. A baker does not consciously set out to serve society by baking bread - he only intends to make money for himself; but natural economic forces lead him to perform a public service, since if he were not doing something useful, people would not pay him for it. Adam Smith expressed this idea in the following way:

"As every individual, therefore, endeavors as much as he can, both to employ his capital in support of domestic industry, and so to direct that industry that its produce may be of greatest value, each individual necessarily labours to render the annual revenue of the Society as great as he can."

"He generally, indeed, neither intends to promote the public interest, nor knows how much he is promoting it. By preferring the support of domestic to that of foreign industry, he intends only his own security; and by directing that industry in such a manner as its produce may be of the greatest value, he intends only his own gain; and he is in this, as in many other cases, led by an invisible hand to promote an end which was no part of his intention. Nor is it always the worse for Society that it was no part of it. By pursuing his own interest, he frequently promotes that of society more effectively than when he really intends to promote it."

In Adam Smith's model of the free market, competition regulates prices. For example, if a baker charges an excessive price for bread, other bakers will undersell him, forcing him to lower his prices if he wants to retain a share of the market.

Not only are prices regulated by the automatic mechanisms of free competition, but also the type and quantity of goods produced: If there are

Fig. 1.5 *Adam Smith (1723-1790).*

too many bakers, producing too much bread, competition will force some of them out of business. Conversely, if the demand for bread greatly exceeds the supply, high prices will attract new tradesmen to the business of baking.

According to Adam Smith, the mechanisms of the free market are capable of organizing the enormously complex economic operations of an advanced civilization, without governmental regulation, driven only by self-interest and guided only by the "invisible hand" of free competition.

"Observe the accommodation of the most common artificer or day labourer in a civilized and thriving country", Smith wrote, "and you will perceive that the number of people of whose industry a part, though but a small part, has been employed in procuring him this accommodation, exceeds all computation. The woolen coat, for example, which covers the day-labourer, as coarse and rough as it may seem, is the joint labour of a great multitude of workmen. The Shepperd, the sorter of wool, the wool-comber, the carder, the dyer, the scribbler, the spinner, the weaver, the fuller, the dresser, with many others, must all join their different arts in order to complete even this homely production. How many merchants and carriers, besides, must have been employed... how much commerce and navigation... how many ship-builders, sailors, sail-makers, rope makers..."

Smith's model is by no means static. The enterprising manufacturer does not spend his profits on luxuries for himself and his family; he reinvests them! He purchases new machinery, starts new factories and new enterprises. In Adam Smith's words, a manufacturer who ignores this commandment to reinvest his profits is "...like him who perverts the revenues of some pious foundation to profane purposes; he pays the wages of idleness with those funds which the frugality of his forefathers had, as it were, consecrated to the maintenance of industry."

With more and more factories, it might seem logical that labor would become scarce and that the wages of factory workers would rise, thus slowing and finally stopping the expansion of the system. Not so, says Smith! When wages rise, the supply of workers will also rise, since more of the children of the poor will survive if they are better fed and better accommodated. Wages will then fall back to the subsistence level. As Smith puts it, "...the demand for men, like that for any other commodity, necessarily regulates the production of men."

This was, in fact, an accurate description of the conditions of the period. Almost everywhere in England and Scotland, only half the children lived to the age of eight or ten. In some places the infant mortality was still higher. "It is not uncommon", Smith wrote, "...in the Highlands of Scotland for a mother who has borne twenty children not to have two alive."

Thus, Smith believed, when the demand for workers forces wages up, the population will increase, and wages will fall again. The expansion of industry will thus continue in an ever-ascending spiral of growth. The question of whether or not economic growth could continue indefinitely did not worry Adam Smith, since he believed that growth could continue as far into the future as it was possible to foresee.

Suggestions for Further Reading

(1) F.K. Brown, *The Life of William Godwin*, J.M. Dent, London, (1926).
(2) P.A. Brown, *The French Revolution in English History*, 2nd edn., Allen and Unwin, London, (1923).

(3) E. Burke, *Reflections on the Revolution in France and on the Proceedings of Certain Societies in London Relative to that Event...*, Dent, London, (1910).
(4) J.B. Bury, *The Idea of Progress*, MacMillan, New York, (1932).
(5) G.G.N. Byron, 6th Baron, *The Works of Lord Byron*, poetry ed. E.H. Coleridge, letters and journals ed. R.E. Prothero, 13 vols., Murray, London, (1898-1904).
(6) K.C. Carter, (ed.), *Enquiry Concerning Political Justice by William Godwin, with Selections from Godwin's other Writings, Abridged and Edited*, Clarendon, Oxford, (1971).
(7) I.R. Christie, *Stress and Stability in Late Eighteenth Century Britain; Reflections on the British Avoidance of Revolution* (Ford Lectures, 1983-4), Clarendon, Oxford, (1984).
(8) M. de Condorcet, *Sur l'admission des Femmes au droit de Cité*, Paris, (1790).
(9) M. de Condorcet, *Esquisse d'un tableau historique des progrès de l'esprit humain*, 2nd edn., Agasse, Paris, (1795).
(10) C.B. Cone, *The English Jacobins: Reformers in late Eighteenth-Century England*, Scribner, New York, (1968).
(11) H.T. Dickenson, *Liberty and Property, Political Ideology in Eighteenth Century Britain*, Holmes and Meier, New York, (1977).
(12) W. Eltis, *The Classical Theory of Economic Growth*, St. Martin's, New York, (1984).
(13) M. Freeman, *Edmund Burke and the Critique of Political Radicalism*, Blackwell, Oxford, (1980).
(14) W. Godwin, *Enquiry Concerning Political Justice and its Influence on Morals and Happiness*, 2 vols. Robinson, London (1793).
(15) W. Godwin, *Things as they Are; or, the Adventures of Calib Williams*, 3 vols., Crosby, London, 1794.
(16) W. Godwin, *The Enquirer, Reflections on Education, Manners and Literature in a Series of Essays*, Robinson, London, (1797).
(17) W. Godwin, *Memoirs of the Author of a Vindication of the Rights of Woman*, Johnson, London, (1798).
(18) W. Godwin, *Thoughts Occasioned by a Perusal of Dr Parr's Spital Sermon...*, Johnson, London, (1801).
(19) E. Halévy, *A History of the English People in the Nineteenth Century*, (transl. E.I. Watkin), 2nd edn., Benn, London, (1949).
(20) E. Halévy, *The Growth of Philosophic Radicalism*, (transl. M. Morris), new edn., reprinted with corrections, Faber, London, (1952).
(21) W. Hazlitt, *The Complete Works of William Hazlitt*, ed. P.P. Howe, after the edition of A.R. Walker and A. Glover, 21 vols., J.M. Dent, London, (1932).

(22) W. Hazlitt, *A Reply to the Essay on Population by the Rev. T.R. Malthus...*, Longman, Hurst, Rees and Orme, London, (1807).
(23) R. Heilbroner, *The Worldly Philosophers: The Lives, Times and Ideas of the Great Economic Thinkers*, 5th edn., Simon and Schuster, New York, (1980).
(24) R.K. Kanth, *Political Economy and Laissez-Faire: Economics and Ideology in the Ricardian Era*, Rowman and Littlefield, Totowa N.J., (1986).
(25) J.M. Keynes, *Essays in Biography*, in *The Collected Writings of John Maynard Keynes*, MacMillan, London, (1971-82).
(26) F. Knight, *University Rebel: The Life of William Frend, 1757-1841*, Gollancz, London (1971).
(27) M. Lamb, and C. Lamb, *The Works of Charles and Mary Lamb*, ed. E.V. Lucas, 7 vols., Methuen, London, (1903).
(28) A. Lincoln, *Some Political and Social Ideas of English Dissent, 1763-1800*, Cambridge University Press, (1938).
(29) D. Locke, *A Fantasy of Reason: The Life and Thought of William Godwin*, Routledge, London, (1980).
(30) J. Locke, *Two Treatises on Government. A Critical Edition with an Introduction and Apparatus Criticus*, ed. P. Laslett, Cambridge University Press, (1967).
(31) J. Macintosh, *Vindicae Gallicae. Defense of the French Revolution and its English Admirers against the Accusations of the Right Hon. Edmund Burke...*, Robinson, London, (1791).
(32) J. Macintosh, *A Discourse on the Study of the Law of Nature and of Nations*, Caldell, London, (1799).
(33) T. Paine, *The Rights of Man: being an Answer to Mr. Burke's Attack on The French Revolution*, Jordan, London, part I (1791), part II (1792).
(34) P.B. Shelley, *The Complete Works of Percy Bysshe Shelley*, ed. R. Ingpen and W.E. Peck, 10 vols., Benn, London; Scribner, New York, (1926-30).
(35) A. Smith, *The Theory of Moral Sentiments...* (1759), ed. D.D. Raphael and A.L. MacPhie, Clarendon, Oxford, (1976).
(36) A. Smith, *An Inquiry into the Nature and Causes of the Wealth of Nations* (1776), Everyman edn., 2 vols., Dent, London, (1910).
(37) H.G. Wells, *Anticipations of the Reaction of Mechanical and Scientific Progress on Human Life and Thought*, Chapman and Hall, London, (1902).
(38) B. Wiley, *The Eighteenth Century Background: Studies of the Idea of Nature in the Thought of the Period*, Chatto and Windus, London, (1940).

(39) M. Wollstonecraft, *A Vindication of the Rights of Men, in a Letter to the Right Honourable Edmund Burke; occasioned by his Reflections on the Revolution in France*, 2nd edn., Johnson, London, (1790).
(40) W. Wordsworth, *Poetical Works*, ed. T. Hutchinson; new edn., rev. E. de Selencourt, Oxford University Press, (1950).
(41) W. Wordsworth, *The Prose Works of William Wordsworth*, eds. W.J.B. Owen and J.W. Smyser, 3 vols., Oxford, (1974).
(42) A. Smith, *The Theory of Moral Sentiments*, first published 1759, 6th Edition, A. Millar, London, (1790).
(43) A. Smith, *An Enquery into the Nature and Causes of the Wealth of Nations*, first published 1776, 5th Edition edited by Edwin Cannan, Methuen, London, (1904).
(44) J. Bonar, *The Theory of Moral Sentiments by Adam Smith*, Journal of Philosophical Studies, **1**, 333-353, (1926).
(45) G.R. Morrow, *The Ethical and Economic Theories of Adam Smith: A Study in the Social Philosophy of the 18th Century*, Cornell Studies in Philosophy, **13**, 91-107, (1923).
(46) H.W. Schneider, ed., *Adam Smith's Moral and Political Philosophy*, Harper Torchbook edition, New York, (1948).
(47) F. Rosen, *Classical Utilitarianism from Hume to Mill*, Routledge, (2003).
(48) J.Z. Muller, *The Mind and the Market: Capitalism in Western Thought*, Anchor Books, (2002).
(49) J.Z. Muller, *Adam Smith in His Time and Ours: Designing the Decent Society*, Princeton University Press, (1995).
(50) S. Hollander, *The Economics of Adam Smith*, University of Toronto Press, (19773).
(51) K. Haakonssen, *The Cambridge Companion to Adam Smith*, Cambridge University Press, (2006).
(52) K. Haakonssen, *The Science of a Legeslator: The Natural Jurisprudence of David Hume and Adam Smith*, Cambridge University Press, (1981).
(53) I. Hont and M. Ignatieff, *Wealth and Virtue: The Shaping of Political Economy in the Scottish Enlightenment*, Cambridge University Press, (1983).
(54) I.S. Ross, *The Life of Adam Smith*, Clarendon Press, Oxford, (1976).
(55) D. Winch, *Adam Smith's Politics: An Essay in Historiographic Revision*, Cambridge University Press, (1979).

Chapter 2

THE DISMAL SCIENCE

2.1 Malthus

In 1798, an anonymous author published a small book with the title *An Essay on the Principle of Population, as It Affects the Future Improvement of Society, with Remarks on the Speculations of Mr. Godwin, M. Condorcet, and Other Writers.* It was the outcome of conversations between Daniel Malthus, an intellectual English country gentleman, and his son, Robert.

Daniel Malthus was an ardent admirer of Rousseau and other optimistic philosophers of the Enlightenment, some of whom whom he knew personally. His son, Thomas Robert Malthus (always called "Robert" or "Bob"), had graduated from Cambridge in 1788, as Ninth Wrangler, i.e. the ninth best mathematician in the graduating class. He was the only student in his college to obtain honors in mathematics.

Robert Malthus was a younger son, and therefore not due to inherit his father's estate. Like many others in his situation, he had become a Curate in the Anglican Church. He had been assigned to Oakwood, a parish in a poor woodland district of Surrey. During his work at Oakwood, Robert Malthus became familiar with with the harsh lives of his parishioners. They were almost completely illiterate, and lived in low huts with dirt floors and tiny windows, the walls being made of woven branches plastered with clay. They lived almost entirely on bread. Few of the women and children had shoes or stockings. Growing up under these conditions, the children were stunted and developed late.

It was one of Robert Malthus duties to keep the record of births and deaths in the parish. To his surprise, he noticed that despite the extremely harsh conditions under which his parishioners lived, the number of births was higher than the number of deaths. It was this remarkable fact that first turned his attention to the statistics of population.

Daniel Malthus and his family had settled permanently at Albury, nine miles from Okewood, and it was here that his famous conversations between father and son took place. 1793, the year of Robert's appointment to Okewood, was also the year in which Godwin's *Political Justice* was published. We can imagine the enthusiasm with which Daniel Malthus read Godwin's book and discussed its Utopian ideas with his son.

As Daniel Malthus talked warmly of human progress, Robert's mind turned to the unbalance between births and deaths which he had noticed at Okewood; and he pointed out to his father that no matter what benefits science might be able to confer, they would soon be eaten up by population growth. Regardless of technical progress, the condition of the lowest social class would remain exactly the same: The poor would continue to live, as they always had, on the exact borderline between survival and famine.

For the very poor, Robert told his father, extreme hunger would not last long, since more of their children would then starve, and their numbers would decrease until the food supply became adequate. But change for the better was also impossible, since if more food should become available, more children would then survive, and the population would increase until it once again faced starvation. Robert Malthus realized that this was a very sombre description of the condition of the poor; but observation of his parishioners at Oakwood had convinced him that the gloomy picture was realistic. The lowest class of society seemed to be clinging to a miserable existence on the precise boundary between survival and starvation.

Father and son continued to discuss these issues. Finally Daniel Malthus was so impressed with his son's arguments that he encouraged him to write them down. The result was Robert Malthus' first essay on population. It was only 50,000 words in length, and was published anonymously in 1798. His fundamental idea was that "the power of population is indefinitely greater than the power in the earth to produce subsistence for man".

"That population cannot increase without the means of subsistence", Robert Malthus wrote, "is a proposition so evident that it needs no illustration. That population does invariably increase, where there are means of subsistence, the history of every people who have ever existed will abundantly prove. And that the superior power cannot be checked without producing misery and vice, the ample portion of these two bitter ingredients in the cup of human life, and the continuance of the physical causes that seem to have produced them, bear too convincing a testimony."

Robert Malthus had been trained in mathematics at Cambridge University, and to illustrate the contrast between unrestrained population growth and the possible growth of the food supply, he contrasted the geometrical progression, 1,2,4,8,16,32,64,128,256,..., with the arithmetical progression, 1,2,3,4,5,6,7,8,... (Today we would say that population has a tendency to

Fig. 2.1 *Thomas Robert Malthus (1766-1834).*

grow exponentially with time, while the food supply can at best only grow linearly).

The natural unrestrained growth of population is geometrical (i.e., exponential), Malthus argued. But, exponential demographic growth is seldom actually observed. Therefore, at almost all times, there must be strong forces, such as famine, disease and war, which restrain the growth of population. Occasionally opening of new lands, or the development of new agricultural techniques allow a period of unrestrained growth; but this is rare, and when it happens, a new equilibrium is quickly established. At almost all periods of history, human population presses painfully against the barrier imposed by a limited food supply, or is held in check by other strong forces.

In 1793, when Godwin's *Political Justice* was published, England's mood had been optimistic; but in 1798, when Malthus' *Essay on Population* came out, hopes for reform had been replaced by reaction and pessimism. The change had been caused by Robespierre's Reign of Terror and by the threat of a French invasion. Thus both books came out at exactly the right moment to catch the prevailing mood. In addition to being published at exactly the right moment, Malthus' *Essay* was clearly and powerfully written. Readers were also struck by the contrast between geometrical and arithmetical progressions.

It was clear to William Godwin that Malthus' *Essay on the Principle of Population* was a serious argument against the Utopian ideas of *Political Justice.* Although the *Essay* had been published anonymously, Godwin soon found out the name of its author, and on several occasions he invited Malthus to his home for breakfast and for discussions of political and economic problems.

In 1801, Godwin published a reply to his various critics. In replying to Malthus, Godwin granted that the problem of overpopulation was an extremely serious one. However, Godwin wrote, all that is needed to solve the problem is a change of the attitudes of society. For example we need to abandon the belief "that it is the first duty of princes to watch for (i.e. encourage) the multiplication of their subjects, and that a man or woman who passes the term of life in a condition of celibacy is to be considered as having failed to discharge the principal obligations owed to the community.

"On the contrary", Godwin wrote, "it now appears to be rather the man who rears a numerous family that has to some degree transgressed the consideration he owes to the public welfare." Godwin believed that once the reasons for it were understood, people would be willing to accept limitations on the size of their families as an obligation to society.

Meanwhile, Robert Malthus began collecting data for a new edition of his book. He realized that his small *Essay* had caught the attention of many thoughtful people in England. However, he now wished to show the points that he had made to be valid not only for one time and place, but for all societies and at all periods of human history. For this purpose, he visited both Scandinavia and Russia, keeping careful notebooks. During the summer of 1802, the Peace of Amiens allowed him to travel on the Continent, and he visited both France and Switzerland.

While working on his second edition, Malthus lived in a garret in London, surrounded by piles of books. He felt that to make a real impact he needed to buttress his thesis with as many facts as possible. Malthus studied the books of great explorers, such as those of Cook, Vancouver, Robertson and Bruce. He also used writings by missionaries, diplomats and traders, as well as histories of ancient Greece and Rome. Additional data was supplied by his own careful notes on conditions in Scandinavia, Russia, Germany, France and Switzerland.

Malthus second edition was more than three times the length of his original essay on population. Its publication in 1803 created a storm, since it dealt with politically sensitive issues. Malthus suddenly found himself at the center of a hotly contested debate. He was denounced by one side and applauded by the other.

Malthus had carefully assembled a massive amount of demographic data, which he used in Books I and II of his 1803 edition. He carefully described the checks to population growth that had operated in all the countries and all the historical periods of which he had knowledge. Malthus begins by describing the enormous potential power of population growth contrasted the slow growth of the food supply. He concludes that strong checks to the increase of population must be operating almost continuously to keep human numbers within the bounds of available sustenance.

Malthus then discusses the types of checks, which may be either "preventative" or "positive". The preventative checks are those which reduce the birth rate, while the positive checks are those which increase the death rate. As examples of positive checks, which increase mortality, he lists "unwholesome occupations, severe labor and exposure to the seasons, extreme poverty, bad nursing of children, great towns, excesses of all kinds, the whole train of common diseases and epidemics, wars, plague, and famine."

Malthus then begins to describe the various societies on which he had collected data. He starts with hunter-gatherer societies, such as the North American Indians, and the inhabitants of New Holland, Van Diemens Land and Tierra del Fuego:

"The great extent of territory required for the support of the hunter has been repeatedly stated and acknowledged", Malthus says, "...The tribes of hunters, like beasts of prey, whom they resemble in their mode of subsistence, will consequently be thinly scattered over the surface of the earth. Like beasts of prey, they must either drive away or fly from every rival, and be engaged in perpetual contests with each other...The neighboring nations live in a perpetual state of hostility with each other. The very act of increasing in one tribe must be an act of aggression against its neighbors, as a larger range of territory will be necessary to support its increased numbers.

The contest will in this case continue, either till the equilibrium is restored by mutual losses, or till the weaker party is exterminated or driven from its country... Their object in battle is not conquest but destruction. The life of the victor depends on the death of the enemy."

Malthus concludes that among hunter-gatherer societies, war is among the main checks to population growth, although infanticide, disease and famine also play a role. He quotes Captain Cook's description of the state of perpetual war that existed among the inhabitants of the region near Queen Charlotte's Sound in New Zealand: "If I had followed the advice of all our pretended friends", Cook had written, "I might have extirpated the whole race; for the people of each hamlet or village, by turns, applied to me to destroy the other." According to Cook, these people not only practiced ceaseless war, but also cannibalism.

Malthus also discusses nomadic societies of the Near East and Asia, and here again, war again appears as one of the major mechanisms by which human numbers are kept in equilibrium with the food supply.

In describing the Germanic tribes of Northern Europe, Malthus quotes Machiavelli's *History of Florence*: "The people who inhabit the northern parts that lie between the Rhine and the Danube", Machiavelli wrote, "living in a healthful and prolific climate, often increase to such a degree that vast numbers of them are forced to leave their native country and go in search of new habitations. When any of those provinces begins to grow too populous and wants to dis-burden itself, the following method is observed. In the first place, it is divided into three parts, in each of which there is an equal portion of the nobility and commonality, the rich and the poor. After this they cast lots; and that division on which the lot falls quits the country and goes to seek its fortune, leaving the other two more room and liberty to enjoy their possessions at home. These emigrations proved the destruction of the Roman Empire."[1]

Malthus then turns to his own time and discusses the nations of Europe. Here positive checks, such as poverty, unsanitary housing, child labor, malnutrition and disease all took a heavy toll, but war was a less important check in 18th century Europe than in hunting and pastoral societies. On the other hand, the preventative checks to population growth, such as late marriage, played a much larger roll.

In Book III of his 1803 edition, Robert Malthus discusses the Utopian ideas put forward by Condorcet and Godwin. "Must not a period then

[1]In the societies that Malthus describes, we can see a clear link not only between population pressure and poverty, but also between population pressure and war. Perhaps this is why the suffering produced by poverty and war saturates so much of human history. Does stabilization of population through birth control offer a key to eliminating this suffering?

arrive", Condorcet had written, "... when the increase of the number of men surpassing their means of subsistence, the necessary result must be either a continual diminution of happiness and population... or at least a kind of oscillation between good and evil?"

Condorcet thought that population pressure would begin to produce suffering only in the very distant future, but Malthus disagrees with him on exactly this point. "M. Condorcet's picture of what may be expected to happen when the number of men shall surpass subsistence is justly drawn... The only point in which I differ from M. Condorcet in this description is with regard to the period when it may be applied to the human race... This constantly subsisting cause of periodical misery has existed in most countries ever since we have had any histories of mankind, and continues to exist at the present moment." Here we must certainly agree with Malthus. He has buttressed his assertions with an enormous amount of data. However, his next paragraph is weaker:

"M. Condorcet, however, goes on to say", Malthus continues, "that should the period, which he conceives to be so distant, ever arrive, the human race, and the advocates of the perfectibility of man, need not be alarmed at it. He then proceeds to remove the difficulty in a manner which I profess not to understand. Having observed that the ridiculous prejudices of superstition would by that time have ceased to throw over morals a corrupt and degrading austerity, he alludes either to a promiscuous concubinage, which would prevent breeding, or to something else as unnatural. To remove the difficulty in this way will surely, in the opinion of most men, be to destroy that virtue and purity of manners which the advocates of equality and of the perfectibility of man profess to be the end and object of their views."

When Malthus says "something else as unnatural", he means birth control (some forms of which existed in his time); but he does not really justify his assumption that birth control is immoral. Is prolonged celibacy really preferable to birth control within marriage as a means of preventing excessive population growth? If so, then why? Malthus does not face these questions, although they lie at the very heart of the problem of population. However, we can perhaps forgive him for gliding much too lightly over these central issues if we remember his position as a Curate in the Anglican Church.

Malthus then turns to William Godwin's *Political Justice*: "The system of equality which Mr. Godwin proposes", Malthus writes, "is, on the first view of it, the most beautiful and engaging which has yet appeared. A melioration of society to be produced merely by reason and conviction gives more promise of permanence than than any change effected and maintained by force. The unlimited exercise of private judgment is a doctrine grand and captivating, and has a vast superiority over those systems where every individual is in a manner the slave of the public. The substitution of benevolence, as a master-spring and moving principle of society, instead of self-love, appears at first sight to be a consummation devoutly to be wished. In short, it is impossible to contemplate the whole of this fair picture without emotions of delight and admiration, accompanied with an ardent longing for the period of its accomplishment."

"But alas!" Malthus continues, "That moment can never arrive.... The great error under which Mr. Godwin labors throughout his whole work is the attributing of almost all the vices and misery that prevail in civil society to human institutions. Political regulations and the established administration of property are, with him, the fruitful sources of all evil, the hotbeds of all the crimes that degrade mankind. Were this really a true state of the case, it would not seem a completely hopeless task to remove evil completely from the world; and reason seems to be the proper and adequate instrument for effecting so great a purpose. But the truth is, that though human institutions appear to be, and indeed often are, the obvious and obtrusive causes of much misery in society, they are, in reality, light and superficial in comparison with those deeper-seated causes of evil which result from the laws of nature and the passions of mankind."

Like Condorcet, Godwin was aware that excessive population growth would some day threaten his Utopia; but he believed, just as Condorcet had done, that the threat belonged to the distant future: "Three-fourths of the habitable globe are now uncultivated", Godwin had written, "The parts already cultivated are capable of immeasurable improvement. Myriads of centuries of still increasing population may pass away, and the earth be still found sufficient for the subsistence of all its inhabitants."

Malthus answers this by saying that if Godwin's Utopian society were ever actually established, it would lead to extremely rapid population growth, perhaps with a doubling time as small as 15 years. But even if the doubling time were longer, for example 25 years[2], exponential increase of population would lead to severe food shortages, not in "myriads of centuries", but in only one or two.

As the threat of famine became acute, Malthus points out, self-love would replace benevolence as the main-spring of human action, and Godwin's Utopia would collapse. Because of the force of population growth, "Man cannot live in the midst of plenty. All cannot share alike the bounties of nature. Were there no established administration of property, every man would be obliged to guard with his force his little store. Selfishness would be triumphant. The subjects of contention would be perpetual. Ev-

[2]This was the doubling time observed in the United States.

ery individual would be under constant anxiety about corporal support, and not a single intellect would be left free to expatiate in the field of thought."

In 1803, when Malthus second *Essay on the Principle of Population* was published, the political situation in England was extremely tense. Many people believed that the country was on the verge of a revolution, similar to those that had occurred in America and France. Food shortages and the Poor Laws were sensitive political issues. Thus, the publication of the second *Essay* created a storm of debate. As Malthus' biographer James Bonar put it, "From the first, Malthus was not ignored. For thirty years it rained refutations."

Publications both for and against the *Essay on Population* began to stream from England's authors. Samuel Taylor Coleridge commented: "Is it not lamentable - is it not even marvelous - that the monstrous practical sophism of Malthus should now have gotten complete possession of the leading men of the kingdom! Such an essential lie in morals - such a practical lie in fact as it is too! I solemnly declare that I do not believe that all the heresies and sects and factions which ignorance and the weakness and wickedness of man have ever given birth to, were altogether so disgraceful to man as a Christian, a philosopher, a statesman or citizen, as this abominable tenet."

Among the other reformers who wrote angrily against the Malthusians was the young poet, Percy Bysshe Shelley, Godwin's son-in-law: "Many well-meaning persons", Shelly wrote, " ... would tell me not to make people happy for fear of over-stocking the world... War, vice and misery are undeniably bad; they embrace all that we can conceive of temporal and eternal evil. Are we to be told that these are remedy-less, because the earth would in case of their remedy, be overstocked?"

Shelley also called Malthus a "priest, eunuch, and tyrant", and accused him of proposing that after then poor "have been stript naked by the tax-gatherer and reduced to bread and tea and fourteen hours of hard labor by their masters... the last tie by which Nature holds them to benignant earth (whose plenty is garnered up in the strongholds of their tyrants) is to be divided... They are required to abstain from marrying under penalty of starvation. And it is threatened to deprive them of that property which is as strictly their birthright as a gentleman's land is his birthright... whilst the rich are to be permitted to add as many mouths to consume the products of the poor as they please."

On the other hand, the author and journalist, Harriet Martineau, wrote: "The desire of his [Malthus'] heart and the aim of his work were that domestic virtue and happiness should be placed within the reach of all... He found that a portion of the people were underfed, and that one consequence of this was a fearful mortality among infants; and another consequence the growth of a recklessness among the destitute which caused infanticide, corruption of morals, and at best, marriage between pauper boys and girls; while multitudes of respectable men and women, who paid rates instead of consuming them, were unmarried at forty or never married at all. Prudence as to time of marriage and for making due provision for it was, one would think, a harmless recommendation enough, under the circumstances."

Meanwhile, undisturbed by the furor that he had caused, Malthus pursued a life of quiet scholarship. In 1805, he accepted a position as Professor of History and Political Economy at the East India Company's College at Haileybury, thus becoming the first professor of economics in England, and probably also the first in the world.

2.2 Ricardo and the Iron Law of Wages

Among Malthus' closest friends was the financier David Ricardo (1772-1823). Ricardo had been born into a Jewish family that had moved to London from Portugal. However, at the age of 21 he had broken relations with his family and rejected his orthodox Jewish faith in order to marry a Quaker girl. Ricardo, who had worked with his father on the London Sock Exchange since the age of 14, then proceeded to become a financier in his own right, amassing a fortune worth over a million pounds, in those days an immense sum.

Having read a copy of Adam Smith's *Wealth of Nations*, Ricardo became interested in theoretical economics, and at the age of 37 he began to write about this subject. His articles and books were admired by Malthus, and the two became close friends, although they disagreed on many issues.

Malthus had been brought up as a member of the British landowning class. He valued the beauty of the countryside, and was disturbed by the growth of industrialism. By contrast, Ricardo's sympathies lay with the rising and vigorous class of industrialists. The theory of rent, developed by Ricardo, showed that there is an inevitable conflict between these two classes.

Ricardo's theory of rent dealt with the effect of economic growth on prices, wages and profits. He and Malthus both agreed with Adam Smith's picture of growth: The virtuous industrialist does not spend his profits on luxuries, but instead reinvests them. New factories are built, the demand for workers increases, wages rise, and more workers are "produced" in response to the demand, i.e., more of the worker's children survive, and their numbers grow.

With each turn of the spiral of economic growth, there is an increased demand for food, since the population of workers increases. The most fertile land is already in use, but to meet the larger demand for food, marginal land is tilled, for example land on steep hillside slopes. It costs more to grow grain on marginal land, and therefore grain prices rise. According to Ricardo, the only people who benefit from economic growth are the owners of especially fertile land. The factory owners do not benefit, because they must pay higher wages to meet the increased price of food for their workers, and their profits remain the same. The workers do not benefit, because regardless of the price of grain, each of them is given only enough food to survive. The true beneficiaries of economic growth, according to Ricardo, are the owners of the most fertile land, i.e., the landowning aristocracy.

Ricardo defines "rent" to be the difference, per acre, between the cost of growing grain on good land, and the cost on marginal land. This difference is pocketed by the owners of good land. They do not really deserve it because ownership of fertile land is something that they inherited, rather than something that they produced by their own efforts.

At the time when Ricardo was writing, imports of cheap foreign grain were effectively blocked by the Corn Laws, a series of acts of Parliament which were in force between 1815 and 1846. These laws imposed prohibitively high tariffs on the import of foreign grain. Ricardo's theory of rents showed that the Corn Laws benefited the landowning aristocracy at the expense of the industrialists. His sympathies were with the industrialists, because he felt that the Corn Laws were forcing England back into feudalism and economic stagnation. By contrast, Malthus favored the Corn laws because he felt that it was dangerous for England to become dependent on imports of foreign grain. What would the country do in case of war?, Malthus asked. What would England do if it lost its industrial edge and became unable to export its manufactured products? How would the country then support its overgrown population?

In the end, the aristocracy lost its control of Parliament, the Corn Laws were repealed, and the population of England continued to grow. It has grown from 8.3 million in 1801, the year of the first census, to 50.7 million in 2006. Today, England could not possibly support its population on home-grown food. Like the Netherlands and Japan, Britain is dependent on exports of manufactured goods and imports of grain.

Ricardo believed that the "natural price" of any commodity is the lowest possible cost of its production, and that in the long run, prices of any commodity would approach this natural value. When he applied this idea to labor, the result was his "Iron Law of Wages". Since the lowest cost of "producing" workers is the cost of keeping them alive at the subsistence level, he reasoned, the natural price of labor is determined by the lowest possible cost of sustenance. If workers are paid less than this, they will die, their numbers will decrease, the demand for workers will increase, and

Fig. 2.2 *David Ricardo (1772-1823).*

the price of labor will rise. If they are paid more, a greater number of their children will survive, the number of workers will increase above demand, and wages will fall. According to this argument, starvation wages are inevitable.

Ricardo's reasoning assumes industrialists to be completely without social conscience or governmental regulation; it fails to anticipate the development of trade unionism; and it assumes that the working population will multiply without restraint as soon as their wages rise above the starvation level. This was an accurate description of what was happening in England during Ricardo's lifetime, but it obviously does not hold for all times and all places.

Fig. 2.3 *Jean-Baptiste Say (1767-1832). He believed that a general glut is impossible, because by producing goods, the producers earn enough money to buy back the aggregate production, no matter how large the total volume of goods may be.*

Malthus worried about the problem of a "general glut", i.e. the failure of an economy to buy back all of its output. Anticipating Hobson, he thought that such a situation might result from too much saving. Ricardo, on the other hand, considered a "general glut" to be impossible. He based his opinion on Say's Law, a proposition that had been developed by the French economist, Jean-Baptiste Say (1767-1832).

Say postulated that the desire for commodities is infinite. Although a person can consume only a finite amount of food, Say thought, the desire for such objects as clothes, houses, furniture, paintings etc. has no limits. Say also maintained that the act of producing goods also generates the purchasing power needed to buy them back. "How else should it be possible that there should now be bought and sold in France five or six times as many commodities as in the miserable reign of Charles VI?", Say asked. James Mill, the father of John Stuart Mill, anticipated Say's Law by stating that "supply creates its own demand".

Malthus believed that although Say's law might hold in the long run, it did not necessarily hold over short intervals of time. Later, in the 20th century, the short-term version of Say's Law was also attacked by John Maynard Keynes.

In the 21st century and in the more distant future, Say's Law will become problematic because the exhaustion of non-renewable resources will impose limits on economic growth. It will then become necessary to have some definition of an upper limit to the commodities required by each of the citizens of a stable future society.

Suggestions for Further Reading

(1) A. Annesley, *Strictures on the True Cause of the Alarming Scarcity of Grain and Other Provisions*, Murray and Highley, London, (1800).
(2) J. Bonar, *Malthus and his Work*, Allen and Unwin, London, (1894); 2nd edn. with notes and expanded biography, Allen and Unwin, London, (1924).
(3) H.A. Boner, *Hungry Generations: the Nineteenth-Century Case Against Malthusianism*, King's Crown Press, Columbia University, New York, (1955).
(4) E. Burke, *Reflections on the Revolution in France and on the Proceedings of Certain Societies in London Relative to that Event...*, Dent, London, (1910).
(5) A.M. Carr-Saunders, *The Population Problem*, Oxford University Press, (1922).
(6) I.R. Christie, *Stress and Stability in Late Eighteenth Century Britain; Reflections on the British Avoidance of Revolution* (Ford Lectures, 1983-4), Clarendon, Oxford, (1984).
(7) H.T. Dickenson, *Liberty and Property, Political Ideology in Eighteenth Century Britain*, Holmes and Meier, New York, (1977).

(8) E. Halévy, *A History of the English People in the Nineteenth Century*, (transl. E.I. Watkin), 2nd edn., Benn, London, (1949).
(9) R. Heilbroner, *The Worldly Philosophers: The Lives, Times and Ideas of the Great Economic Thinkers*, 5th edn., Simon and Schuster, New York, (1980).
(10) G. Himmelfarb, *The Idea of Poverty: England in the Early Industrial Age*, Knopf, New York, (1984).
(11) T.H. Huxley, *Evolution and Ethics and Other Essays*, Appleton, London, n.d.
(12) P. James, *Population Malthus: his Life and Times*, Routledge, London, (1979).
(13) R.K. Kanth, *Political Economy and Laissez-Faire: Economics and Ideology in the Ricardian Era*, Rowman and Littlefield, Totowa N.J., (1986).
(14) F. Knight, *University Rebel: The Life of William Frend, 1757-1841*, Gollancz, London (1971).
(15) T.R. Malthus, *An Essay on the Principle of Population as it Affects the Future Improvement of Society, with Remarks on the Speculations of Mr. Godwin, M. Condorcet, and Other Writers*, Johnson, London, (1798).
(16) T.R. Malthus, *An Essay on the Principle of Population, or, A View of its Past and Present Effects on Human Happiness, with an Inquiry into our Prospects Respecting its Future Removal or Mitigation of the Evils which it Occasions*, 2nd edn., Johnson, London (1803); 6th edn. with an introduction by T.H. Hollingsworth, Everyman's University Library, J.M. Dent and Sons, London, (1973).
(17) T.R. Malthus, *An Inquiry into the Nature and Progress of Rent, and the Principles by which it is Regulated*, Murray, London, (1815).
(18) T.R. Malthus, *The Grounds of an Opinion on the Policy of Restricting the Importation of Foreign Corn...*, Murray, London, (1815).
(19) T.R. Malthus, *A Summary View of the Principle of Population*, Murray, London, (1830).
(20) T.R. Malthus, *Principles of Political Economy, Considered with a View to their Practical Application*, 2nd edn., Pickering, London, (1836).
(21) T.R. Malthus, *The Works of T.R. Malthus*, ed. E.A. Wrigley and D. Souden, 8 vols., Pickering, London, (1836).
(22) G.F. McCleary, *The Malthusian Population Theory*, Faber, London, (1953).
(23) J.S. Mill, *Principles of Political Economy*, A new edn., ed. W.J. Ashley, Longmans, London, (1909).

(24) J.S. Nickerson, *Homage to Malthus*, Kennikat Press, Port Washington, N.Y., (1975).
(25) F. Place, *Illustrations and Proofs of the Principle of Population: Including an Examination of the Proposed Remedies of Mr. Malthus, and a Reply to the Objections of Mr. Godwin and Others*, Longman, Hurst, Rees, Orme and Brown, (1822).
(26) D. Ricardo, *On the Principles of Political Economy*, Murray, London, (1817).
(27) D. Ricardo, *Notes on Malthus' 'Measure of Value'*, ed. Pier Luigi Porta, Cambridge University Press, (1992).
(28) D. Ricardo, *The Works and Correspondence of David Ricardo*, ed. P. Sraffa, 11 vols., Cambridge University Press, (1951-73).
(29) P.B. Shelley, *The Complete Works of Percy Bysshe Shelley*, ed. R. Ingpen and W.E. Peck, 10 vols., Benn, London; Scribner, New York, (1926-30).
(30) K. Smith, *The Malthusian Controversy*, Routledge, London, (1951).
(31) R. Southey, 'Malthus Essay on Population', *Annual Review*, pages 292-301, (1803).
(32) R. Southey, 'Inquiry into the Poor Laws, etc.', *Quarterly Review* VIII, pages 319-56, (December, 1812).
(33) R. Southey, *Essays, Moral and Political*, 2 vols., Murray, London, (1832).
(34) L. Stephen, *History of English Thought in the Eighteenth Century*, 2 vols., Smith, Elder, London, (1876); 2nd edn., (1881).
(35) G.M. Trevelyan, *British History in the Nineteenth Century*, Longmans, London, (1922).
(36) M. Turner, *Malthus and his Time*, MacMillan,, London, (1986).
(37) R. Wallace, *Various Prospects of Mankind, Nature and Providence*, Millar, Edinburgh, (1761).
(38) A.M.C. Waterman, *Revolution, Economics and Religion: Christian Political Economy, 1798-1833*, Cambridge University Press, (1991).
(39) S. Webb and B. Webb, *English Local Government*, 9 vols., Longmans, Green and Co., London, (1906-29).
(40) B. Wiley, *The Eighteenth Century Background: Studies of the Idea of Nature in the Thought of the Period*, Chatto and Windus, London, (1940).
(41) D. Winch, *Malthus*, Oxford University Press, (1987).
(42) A. Young, *The Question of Scarcity Plainly Stated, and Remedies Considered*, B. M'millan, London, (1800).
(43) R.M. Young, 'Malthus and the Evolutionists: the Common Context of Social and Biological Thought', *Past and Present* 43, pages 109-41, (1969).

Chapter 3

DARK SATANIC MILLS

We have seen how the development of printing in Europe produced a brilliant, chainlike series of scientific discoveries. During the 17th century, the rate of scientific progress gathered momentum, and in the 18th and 19th centuries, the practical applications of scientific knowledge revolutionized the methods of production in agriculture and industry.

During the Industrial Revolution, feudal society, with its patterns of village life and its traditional social obligations, was suddenly replaced by a money-dominated society whose rules were purely economic, and in which labor was regarded as a commodity. The changes produced by the industrial revolution at first resulted in social chaos - enormous wealth in some classes of society, and great suffering in other classes; but later, after the appropriate social and political adjustments had been made, the improved methods of production benefited all parts of society in a more even way.

3.1 Development of the steam engine

The discovery of atmospheric pressure

Early steam engines made use of the pressure of the atmosphere, and in fact it was the discovery of atmospheric pressure that led to the invention of the steam engine. Aristotle had maintained "nature abhors a vacuum", but this doctrine was questioned by the Italian physicist Evangelesta Torricelli (1608-1647), who invented the barometer in 1643.

Pump makers working for the Grand Duke of Tuscany had found that suction pumps were unable to raise water to heights greater than 10 meters (in today's units). Attempting to understand why this should be the case, Torricelli filled an approximately 1-meter-long glass tube with mercury, which is 14 times denser than water. The tube was sealed at one end, and open at the other. He then immersed the open end in a dish of mercury, and raised the sealed end, so that the tube was in a vertical position. Part of the mercury flowed out of the tube into the dish, leaving a 76-centimeter-high column of mercury, and 24 centimeters of empty space at the top. The empty space contained what we now call a Torricellian vacuum.

This experiment enabled Torricelli to understand why the Grand Duke's suction pumps were unable to raise water to a height greater than 10 meters. Torricelli realized that both the 10 meter column of water (the maximum that could be achieved), and the (equally heavy) 76 centimeter column of mercury, were held in place by the weight of the atmosphere, which they exactly balanced. Later experiments soon demonstrated that the height of the column of mercury in Torricelli's barometer depended on the weather, and on height above sea level. Summarizing his experiments, Torricelli wrote: "We live submerged at the bottom of an ocean of elementary air, which is known by incontestable experiments to have weight."

Torricelli's experiments marked the start of period where, throughout Europe, much interest was focused on experiments with gases. In 1650 Otto von Guericke, the Mayor of Magdeburg Germany, invented the first vacuum pump. In a dramatic experiment, performed in 1663 in the presence of Frederick Wilhelm I of Brandenburg, von Guericke's assistants fitted two large copper hemispheres together, after the joining surfaces had been carefully greased to make the junction airtight. Von Guericke's pump was then used to evacuate the volume within the hemispheres. To the amazement of the watching crowd, a team of 24 horses, 12 on each side, strained at the hemispheres but failed to separate them. Von Guericke explained that it was the pressure of the atmosphere that held the hemispheres so tightly together, and he demonstrated that when air was allowed to enter the interior volume, the hemispheres could be separated without effort.

Steam engines using atmospheric pressure

Continuing the vogue for experiments with gases and pumps that was sweeping across Europe, Edward Somerset, the 2nd Marquess of Worcester, designed steam-powered pumps to bring water from wells to fountains. He published the designs for his engines in 1663, and he may have installed pumps built according to these designs at Vauxhall House in London. In the 1680's a number of steam-powered pumps were constructed for Louis XIV of France by Sir Samuel Morland (1625-1695), who lived in Vauxhall and may have been influenced by Somerset's ideas.

Meanwhile, in France, the physicist Denis Papin (1647-1712) had become interested in the motive force of steam. Together with Gottfried Leibniz he invented the pressure cooker, and he also invented designs for steam engines. Some of Papin's steam engine designs were presented to the

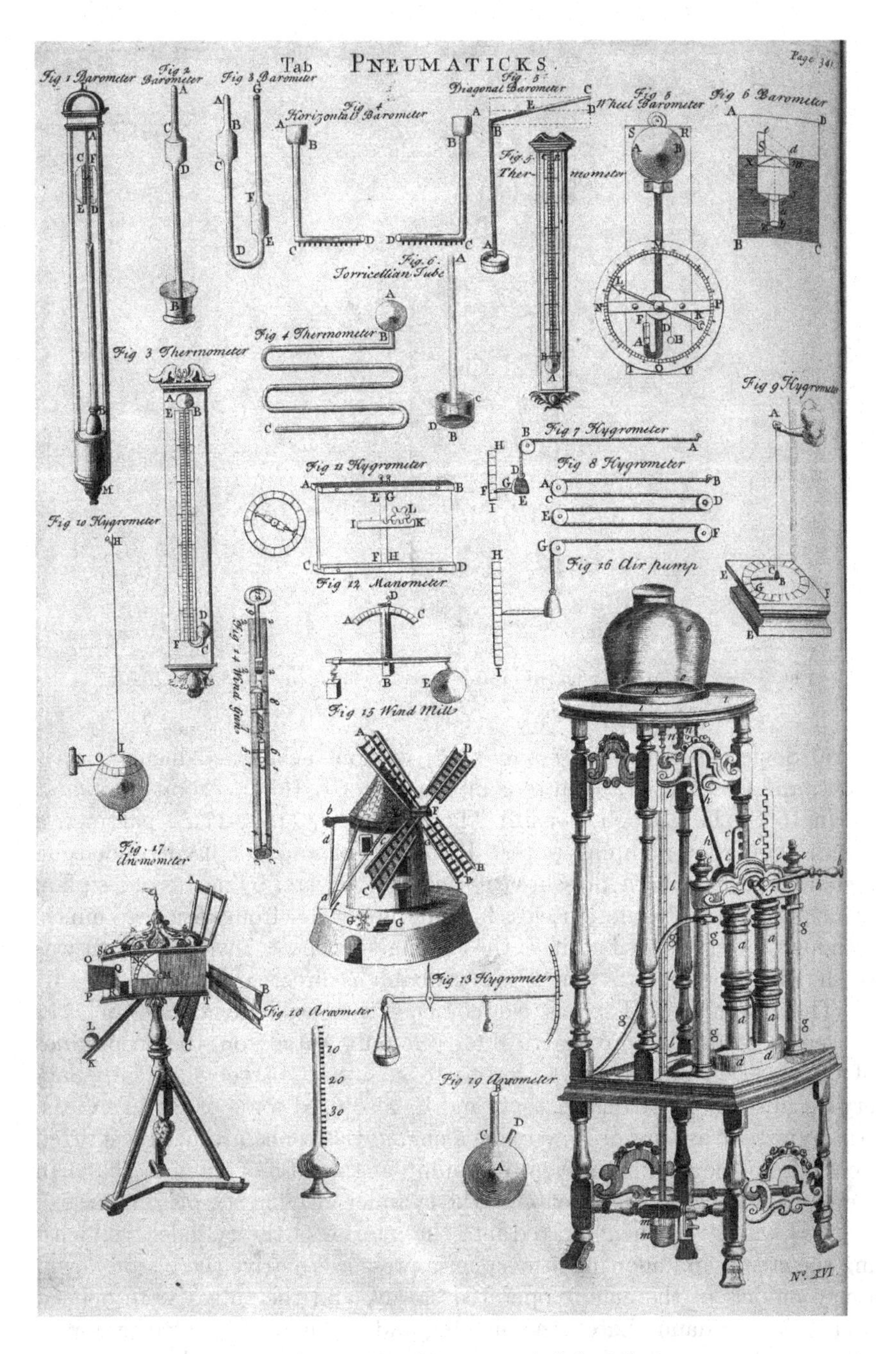

Fig. 3.1 *"Table of Pneumaticks" (1728).*

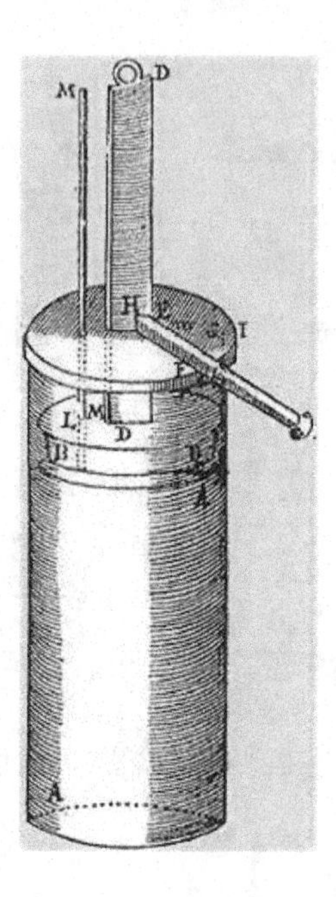

Fig. 3.2 *The French physicist Papin's design for a steam engine (1690).*

Royal Society between 1707 and 1712, without acknowledgment or payment, and this caused Papin to complain bitterly. He died soon afterward.

In 1698, the English inventor Thomas Savery (1650-1715) patented a steam engine for pumping water. It had no piston, but used condensing steam and atmospheric pressure to bring up the water by means of a siphon principle. It was therefore useless for pumping water from very deep mines, although Savery described it as the "Miner's Friend". Savery's design was so similar to Somerset's that it was probably a direct copy.

The ironmonger Thomas Newcomen's "atmospheric-engine" of 1712 proved to be much more practical for pumping water from the deep mines of Cornwall. Newcomen was forced to go into a partnership with Savery because of the latter's patent, and he also used some of Papin's ideas. An important feature of Newcomen's engine was a beam that transmitted power from the working piston to a pump at the base of the mineshaft. In Newcomen's engine, steam entered the cylinder, driving the piston upward. A jet of water was then sprayed into the interior of the cylinder, condensing the steam and allowing atmospheric pressure to drive the piston down. Early models of the engine operated slowly, and the valves were opened and closed by hand. Later, the opening and closing of the valves was performed automatically by means of the "potter cord". According to legend this device is named after a boy, Humphrey Potter, who in 1713 had been given the job of opening and closing the valves. Wishing to play with his friends, he invented the automatic mechanism.

The main problem with Newcomen's engine was that its fuel use was enormously wasteful. This was because, with every cycle, the cylinder was cooled by water, and then heated again by steam.

At Glasgow University, where Adam Smith was Professor of Moral Philosophy, there was a shop where scientific instruments were made and sold. The owner of the shop was a young man named James Watt (1736-1819), who came from a family of ship builders and teachers of mathematics and navigation. Besides being an extremely competent instrument maker, Watt was a self-taught scientist of great ability, and his shop became a meeting place for scientifically inclined students.

James Watt tried to repair the university's small-scale model of the Newcomen engine, but he failed to make it work well. He could see that it was extraordinarily inefficient in its use of fuel, and he began making experiments to find out why it was so wasteful. James Watt quickly found the answer: The engine was inefficient because of the large amounts of energy needed to heat the iron cylinder. In 1765, Watt designed an improved engine with a separate condenser. The working cylinder could then be kept continuously hot.

To have an idea for a new, energy-saving engine was one thing, however, and to make the machine practical was another. James Watt had experience as an instrument maker, but no experience in large-scale engineering. However, Watt formed a partnership with Mathew Boulton, who was the most talented and progressive manufacturer in England.

Boulton was more interested in applying art and science to manufacturing than he was in simply making money. His idea was to bring together under one roof the various parts of the manufacturing process which had been scattered among many small workshops by the introduction of division of labor. He believed that improved working conditions would result in an improved quality of products.

With these ideas in mind, Matthew Boulton built a large mansion-like house on his property at Soho, outside Birmingham, and installed in it all the machinery necessary for the complete production of a variety of small steel products. Because of his personal charm, and because of the comfortable working conditions at the Soho Manufactory, Boulton was able to attract the best and most skillful craftsmen in the region; and by 1765, the number of the staff at Soho had reached 600.

At this point, Erasmus Darwin (the grandfather of Charles Darwin) introduced James Watt to Matthew Boulton, and they formed a partnership for the development of the steam engine. The high quality of craftsmanship

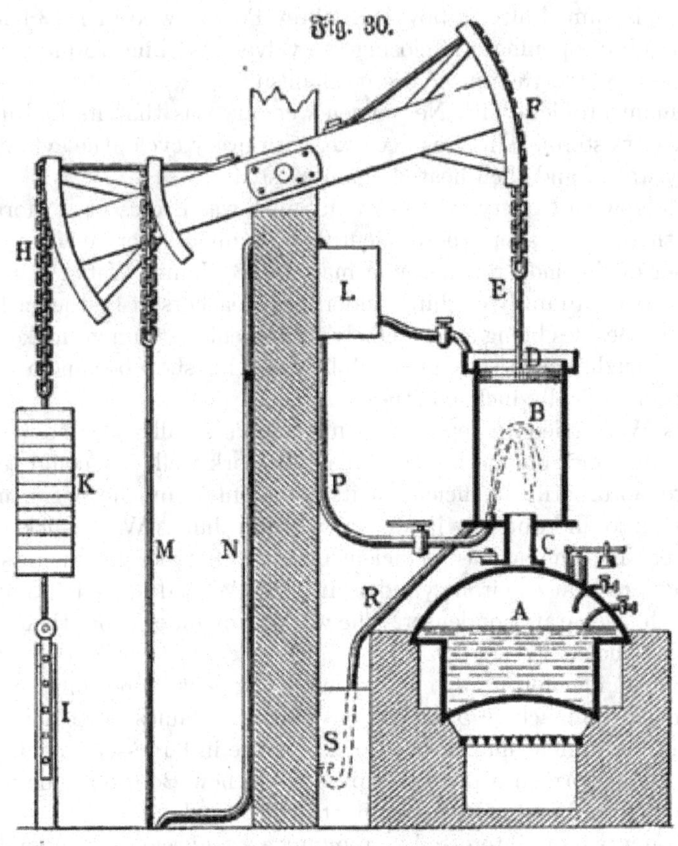

Fig. 3.3 *Newcomen's steam engine.*

and engineering skill which Matthew Boulton was able to put at Watt's disposal allowed the young inventor to turn his great idea into a reality. However, progress was slow, and the original patent was running out.

Boulton skillfully lobbied in Parliament for an extension of the patent and, as James Watt put it, "Mr. Boulton's amiable and friendly character, together with his fame as an ingenious and active manufacturer procured me many and very active friends in both houses of Parliament".

In 1775, the firm of Boulton and Watt was granted an extension of the master steam engine patent until 1800. From a legal and financial standpoint, the way was now clear for the development of the engine; and a major technical difficulty was overcome when the Birmingham ironmaster and cannon-maker, John Wilkinson, invented a method for boring large cylinders accurately by fixing the cutting tool to a very heavy and stable boring shaft.

By 1780, Boulton and Watt had erected 40 engines, about half of which pumped water from the deep Cornish tin mines. Even their early models were at least four times as efficient as the Newcomen engine, and Watt continually improved the design. At Boulton's urging, James Watt designed rotary engines, which could be used for driving mills; and he also invented a governor to regulate the speed of his engines, thus becoming a pioneer of automation. By the time its patent of the separate condenser had run out in 1800, the firm of Boulton and Watt had made 500 engines. After 1800, the rate of production of steam engines became exponential, and when James Watt died in 1819, his inventions had given employment, directly or indirectly, to an estimated two million people.

The Soho manufactory became an almost obligatory stop on any distinguished person's tour of England. Samuel Johnson, for example, wrote that he was received at Soho with great civility; and Boswell, who visited Soho on another occasion, was impressed by "the vastness and contrivance" of the machinery. He wrote that he would never forget Matthew Boulton's words to him as they walked together through the manufactory: "I sell here, Sir, what all the world desires to have - Power!"

3.2 Working conditions

Both Matthew Boulton and James Watt were model employers as well as pioneers of the factory system. Boulton had a pension scheme for his men, and he made every effort to insure that they worked under comfortable conditions. However, when he died in 1809, the firm of Boulton and Watt was taken over by his son, Matthew Robbinson Boulton, in partnership with James Watt Jr. The two sons did not have their fathers' sense of social responsibility; and although they ran the firm very efficiently, they seemed to be more interested in profit-making than in the welfare of their workers.

A still worse employer was Richard Arkwright (1732-1792), who held patents on a series of machines for carding, drawing and spinning silk, cotton, flax and wool. He was a rough, uneducated man, who rose from humble origins to become a multimillionaire by driving himself almost as hard as he drove his workers. Arkwright perfected machines (invented by others) which could make extremely cheap and strong cotton thread; and as a result, a huge cotton manufacturing industry grew up within the space

Fig. 3.4 *Manchester in the 1840's.*

of a few years. The growth of the cotton industry was especially rapid after Arkwright's patent expired in 1785.

Crowds of workers, thrown off the land by the Enclosure Acts and by the Clearances in Scotland, flocked to the towns, seeking work in the new factories[1]. Wages fell to a near-starvation level, hours of work increased, and working conditions deteriorated. Dr. Peter Gaskell, writing in 1833, described the condition of the English mill workers as follows:

"The vast deterioration in personal form which has been brought about in the manufacturing population during the last thirty years... is singularly impressive, and fills the mind with contemplations of a very painful character... Their complexion is sallow and pallid, with a peculiar flatness of feature caused by the want of a proper quantity of adipose substance to cushion out the cheeks. Their stature is low - the average height of men being five feet, six inches... Great numbers of the girls and women walk

[1]During the Highland Clearances, families that had farmed the land for generations were violently forced to leave their houses, which were then burned to prevent return. The land was afterward used as pasturage for sheep, which had been found to be more profitable. Donald McLeod, a crofter (small farmer) in Sutherland, has left the following account of the Clearances in his district: "The consternation and confusion were extreme. Little or no time was given for the removal of persons or property; the people striving to remove the sick and helpless before the fire should reach them; next, struggling to save the most valuable of their effects. The cries of the women and children, the roaring of the affrighted cattle, hunted at the same time by the yelling dogs of the shepherds amid the smoke and fire, altogether presented a scene that completely baffles description - it required to be seen to be believed... The conflagration lasted six days, until the whole of the dwellings were reduced to ashes or smoking ruins."

lamely or awkwardly... Many of the men have but little beard, and that in patches of a few hairs... (They have) a spiritless and dejected air, a sprawling and wide action of the legs..."

"Rising at or before daybreak, between four and five o'clock the year round, they swallow a hasty meal or hurry to the mill without taking any food whatever... At twelve o'clock the engine stops, and an hour is given for dinner... Again they are closely immured from one o'clock till eight or nine, with the exception of twenty minutes, this being allowed for tea. During the whole of this long period, they are actively and unremittingly engaged in a crowded room at an elevated temperature."

Dr. Gaskell described the housing of the workers as follows:

"One of the circumstances in which they are especially defective is that of drainage and water-closets. Whole ranges of these houses are either totally undrained, or very partially... The whole of the washings and filth from these consequently are thrown into the front or back street, which, often being unpaved and cut into deep ruts, allows them to collect into stinking and stagnant pools; while fifty, or even more than that number, having only a single convenience common to them all, it is in a very short time choked with excrementous matter. No alternative is left to the inhabitants but adding this to the already defiled street."

"It frequently happens that one tenement is held by several families... The demoralizing effects of this utter absence of domestic privacy must be seen before they can be thoroughly appreciated. By laying bare all the wants and actions of the sexes, it strips them of outward regard for decency - modesty is annihilated - the father and the mother, the brother and the sister, the male and female lodger, do not scruple to commit acts in front of each other which even the savage keeps hid from his fellows."

"Most of these houses have cellars beneath them, occupied - if it is possible to find a lower class - by a still lower class than those living above them."

The following extract from John Fielden's book, *The Curse of the Factory System* (1836), describes the condition of young children working in the cotton industry:

"It is well known that Arkwright's (so called at least) inventions took manufactures out of the cottages and farmhouses of England... and assembled them in the counties of Derbyshire, Nottinghamshire and more particularly, in Lancashire, where the newly-invented machinery was used in large factories built on the side of streams capable of turning the water wheel. Thousands of hands were suddenly required in these places, remote from towns."

"The small and nimble fingers of children being by far the most in request, the custom instantly sprang up of procuring 'apprentices' from

Fig. 3.5 *London during the industrial revolution.*

the different parish workhouses of London, Birmingham and elsewhere... Overseers were appointed to see to the works, whose interest it was to work the children to the utmost, because their pay was in proportion to the quantity of work which they could exact."

"Cruelty was, of course, the consequence; and there is abundant evidence on record to show that in many of the manufacturing districts, the most heart-rending cruelties were practiced on the unoffending and friendless creatures... that they were flogged, fettered and tortured in the most exquisite refinement of cruelty, that they were, in many cases, starved to the bone while flogged to their work, and that even in some instances they were driven to commit suicide... The profits of manufacture were enormous; but this only whetted the appetite it should have satisfied."

The misery of factory workers in England during the early phases of the Industrial Revolution prompted the writings of Karl Marx (1818-1883) and Frederich Engels (1820-1895). Engels' book, *The condition of the Working Class in England*, was published in 1844. *The Communist Manifesto*, (*Manifest der Komunistischen Partei*), on which Marx and Engels collabo-

Fig. 3.6 *A child working in a South Carolina mill in 1908.*

rated, was published in 1848, while Marx's large book, *Das Kapital. Kritik der politischen Oekonomie* was printed in 1867.

One of the arguments which was used to justify the abuse of labor was that the alternative was starvation. The population of Europe had begun to grow rapidly for a variety of reasons: - because of the application of scientific knowledge to the prevention of disease; because the potato had been introduced into the diet of the poor; and because bubonic plague had become less frequent after the black rat had been replaced by the brown rat, accidentally imported from Asia.

It was argued that the excess population could not be supported unless workers were employed in the mills and factories to produce manufactured goods, which could be exchanged for imported food. In order for the manufactured goods to be competitive, the labor which produced them had to be cheap: hence the abuses. (At least, this is what was argued).

3.3 The slow acceptance of birth control in England

Industrialization benefited England, but in a very uneven way, producing great wealth for some parts of society, but also extreme misery in other

Fig. 3.7 *Child coal miners in Gary, West Virginia.*

social classes. For many, technical progress by no means led to an increase of happiness. The persistence of terrible poverty in 19th-century England, and the combined pessimism of Ricardo and Malthus, caused Thomas Carlyle to call economics "the Dismal Science".

Fortunately, Ricardo's "Iron Law of Wages" seems to have rusted over the years. Apparently it was not an eternal law, but only a description of a passing phase of industrialism, before the appropriate social and legislative adjustments had been made. Among the changes which were needed to insure that the effects of technical progress became beneficial rather than harmful, the most important were the abolition of child labor, the development of unions, the minimum wage law, and the introduction of birth control.

Francis Place (1771-1854), a close friend of William Godwin and James Mill, was one of the earliest and most courageous pioneers of these needed changes. Place had known extreme poverty as a child, but he had risen to become a successful businessman and a leader of the trade union movement.

Place and Mill were Utilitarians, and like other members of this movement they accepted the demographic studies of Malthus while disagreeing with Malthus' rejection of birth control. They reasoned that since abortion and infanticide were already widely used by the poor to limit the size of their families, it was an indication that reliable and humane methods of birth control would be welcome. If marriage could be freed from the miseries which resulted from excessive numbers of children, the Utilitarians believed, prostitution would become less common, and the health and happiness of women would be improved.

Francis Place and James Mill decided that educational efforts would be needed to make the available methods of birth control more widely known and accepted. In 1818, Mill cautiously wrote "The great problem of a real check to population growth has been miserably evaded by all those who have meddled with the subject... And yet, if the superstitions of the nursery were discarded, and the principle of utility kept steadily in view, a solution might not be very difficult to be found."

A few years later, Mill dared to be slightly more explicit: "The result to be aimed at", he wrote in his *Elements of Political Economy* (1821), "is to secure to the great body of the people all the happiness which is capable of being derived from the matrimonial union, (while) preventing the evils which the too rapid increase of their numbers would entail. The progress of legislation, the improvement of the education of the people, and the decay of superstition will, in time, it may be hoped, accomplish the difficult task of reconciling these important objects."

In 1822, Francis Place took the considerable risk of publishing a four-page pamphlet entitled *To the Married of Both Sexes of the Working People*, which contained the following passages:

"It is a great truth, often told and never denied, that when there are too many working people in any trade or manufacture, they are worse paid than they ought to be paid, and are compelled to work more hours than they ought to work. When the number of working people in any trade or manufacture has for some years been too great, wages are reduced very low, and the working people become little better than slaves."

"When wages have thus been reduced to a very small sum, working people can no longer maintain their children as all good and respectable people wish to maintain their children, but are compelled to neglect them; - to send them to different employments; - to Mills and Manufactories, at a very early age. The miseries of these poor children cannot be described, and need not be described to you, who witness them and deplore them every day of your lives."

"The sickness of yourselves and your children, the privation and pain and premature death of those you love but cannot cherish as you wish, need only be alluded to. You know all these evils too well."

"And what, you will ask, is the remedy? How are we to avoid these miseries? The answer is short and plain: the means are easy. Do as other people do, to avoid having more children than they wish to have, and can easily maintain."

"What is to be done is this. A piece of soft sponge is tied by a bobbin or penny ribbon, and inserted just before the sexual intercourse takes place, and is withdrawn again as soon as it has taken place. Many tie a sponge to each end of the ribbon, and they take care not to use the same sponge again until it has been washed. If the sponge be large enough, that is, as large as a green walnut, or a small apple, it will prevent conception... without diminishing the pleasures of married life..."

Fig. 3.8 *Annie Besant (1847-1933).*

"You cannot fail to see that this address is intended solely for your good. It is quite impossible that those who address you can receive any benefit from it, beyond the satisfaction which every benevolent person and true Christian, must feel, at seeing you comfortable, healthy and happy."

The publication of Place's pamphlet in 1822 was a landmark in the battle for the acceptance of birth control in England. Another important step was taken in 1832, when a small book entitled *The Fruits of Philosophy or, the Private Companion of Young Married People* was published by a Boston physician named Dr. Charles Knowlton. The book contained simple contraceptive advice. It reviewed the various methods of birth control available at the time. In order for the sponge method to be reliable, Knowlton's book pointed out, use of a saline douching solution was necessary.

For a number of years, a reprinted edition of Knowlton's book was sold openly in London. However, in 1876 a new law against obscene publications was passed, and a bookseller was sentenced to two year's imprisonment for selling *The Fruits of Philosophy*. Charles Bradlaugh, a liberal politician and editor, and his friend, the feminist author Mrs. Annie Besant, then decided to sell the book themselves in order to provoke a new trial. The Chief Clerk of the Magistrates, the Detective Department, and to the City Solicitor, were all politely informed of the time and place where Charles Bradlaugh and Annie Besant intended to sell Knowlton's book, and the two reformers asked to be arrested.

In the historic trial that followed, the arguments of Malthus were used, not only by Charles Bradlaugh, who conducted his own defense, but also by the Lord Chief Justice, who instructed the jury to acquit the defendants. In the end, the jury ruled that the motives of Besant and Bradley were above reproach. However, the issue was made less clear when the jury also ruled Knowlton's book to be obscene. The enormous publicity that accompanied the trial certainly did not harm the sales of the book!

As birth control was gradually accepted in England, the average number of children per marriage fell from 6.16 in the 1860's to 4.13 in the 1890's. By 1915 the figure had fallen to 2.43. At the same time, trade unions developed, and improved social legislation was enacted. For all of these reasons, conditions improved for the English workers.

Suggestions for further reading

(1) W. Bowden, *Industrial Society in England Towards the End of the Eighteenth Century*, MacMillan, New York, (1925).
(2) G.D. Cole, *A Short History of the British Working Class Movement*, MacMillan, New York, (1927).

(3) P. Deane, *The First Industrial Revolution*, Cambridge University Press, (1969).
(4) Marie Boaz, *Robert Boyle and Seventeenth Century Chemistry*, Cambridge University Press (1958).
(5) J.G. Crowther, *Scientists of the Industrial Revolution*, The Cresset Press, London (1962).
(6) R.E. Schofield, *The Lunar Society of Birmingham*, Oxford University Press (1963).
(7) L.T.C. Rolt, *Isambard Kingdom Brunel*, Arrow Books, London (1961).
(8) J.D. Bernal, *Science in History*, Penguin Books Ltd. (1969).
(9) Bertrand Russell, *The Impact of Science on Society*, Unwin Books, London (1952).
(10) Wilbert E. Moore, *The Impact of Industry*, Prentice Hall (1965).
(11) Charles Morazé, *The Nineteenth Century*, George Allen and Unwin Ltd., London (1976).
(12) Carlo M. Cipolla (editor), *The Fontana Economic History of Europe*, Fontana/Collins, Glasgow (1977).
(13) Martin Gerhard Geisbrecht, *The Evolution of Economic Society*, W.H. Freeman and Co. (1972).
(14) P.N. Stearns, *The Industrial Revolution in World History*, Westvieiw Press, (1998).
(15) E.P. Thompson, *The Making of the English Working Class*, Pennguin Books, London, (1980).
(16) N.J. Smelser, *Social Change and the Industrial Revolution: An Application of Theory to the British Cotton Industry*, University of Chicago Press, (1959).
(17) D.S. Landes, *The Unbound Prometheus: Technical Change and Industrial Development in Western Europe from 1750 to the Present, 2nd ed.*, Cambridge University Press, (2003).
(18) S. Pollard, *Peaceful Conquest: The Industrialization of Europe, 1760-1970*, Oxford University Press, (1981).
(19) M. Kranzberg and C.W. Pursell, Jr., eds., *Technology in Western Civilization*, Oxford University Press, (1981).
(20) M.J. Daunton, *Progress and Poverty: An Economic and Social History of Britain, 1700-1850*, Oxford University Press, (1990).
(21) L.R. Berlanstein, *The Industrial Revolution and Work in 19th Century Europe*, Routledge, (1992).
(22) J.D. Bernal, *Science and Industry in the 19th Century*, Indiana University Press, Bloomington, (1970).

Chapter 4

HOBSON'S THEORY

4.1 The colonial era

The rapid development of technology in the Europe also opened an enormous gap in military strength between the industrialized nations and the rest of the world. Taking advantage of their superior weaponry, the advanced industrial nations rapidly carved the remainder of the world into colonies, which acted as sources of raw materials and food, and as markets for manufactured goods.

Throughout the American continent, the native Indian population had proved vulnerable to European diseases, such as smallpox, and large numbers of them had died. The remaining Indians were driven westward by streams of immigrants arriving from Europe.

Fig. 4.1 *The Aztec emperor, Montezuma II, welcomed Hernando Cortez, a Spanish conquistador, believing him to be a god. The conquistadors were primarily interested in taking the Aztec's gold, which they did with great cruelty. European diseases decimated the indigenous people of both North and South America.*

In the 18th and 19th centuries, the continually accelerating development of science and science-based industry began to affect the whole world. As the factories of Europe poured out cheap manufactured goods, a change took place in the patterns of world trade: Before the Industrial Revolution, trade routes to Asia had brought Asian spices, textiles and luxury goods to Europe. For example, cotton cloth and fine textiles, woven in India, were imported to England. With the invention of spinning and weaving machines, the trade was reversed. Cheap cotton cloth, manufactured in England, began to be sold in India, and the Indian textile industry withered, just as the hand-loom industry in England itself had done a century before.

Often the industrialized nations made their will felt by means of naval bombardments: In 1854, Commodore Perry forced Japan to accept foreign traders by threatening to bombard Tokyo. In 1856, British warships bombarded Canton in China to punish acts of violence against Europeans living in the city. In 1864, a force of European and American warships bombarded Choshu in Japan, causing a revolution. In 1882, Alexandria was bombarded, and in 1896, Zanzibar.

Much that was beautiful and valuable was lost, as mature traditional cultures collapsed, overcome by the power and temptations of modern industrial civilization. For the Europeans and Americans of the late 19th century and early 20th century, progress was a religion, and imperialism was its crusade.

Between 1800 and 1875, the percentage of the earth's surface under European rule increased from 35 percent to 67 percent. In the period between 1875 and 1914, there was a new wave of colonial expansion, and the fraction of the earth's surface under the domination of colonial powers (Europe, the United States and Japan) increased to 85 percent, if former colonies are included.

The unequal (and unfair) contest between the industrialized countries, armed with modern weapons, and the traditional cultures with their much more primitive arms, was summarized by the English poet Hilaire Belloc in a cynical couplet:

Whatever happens, we have got
The Maxim gun, and they have not.

During the period between 1880 and 1914, British industrial and colonial dominance began to be challenged. Industrialism had spread from Britain to Belgium, Germany and the United States, and, to a lesser extent, to France, Italy, Russia and Japan. By 1914, Germany was producing twice as much steel as Britain, and the United States was producing four times as much.

Fig. 4.2 *The Maxim gun was one of the world's first automatic machine guns. It was invented in the United States in 1884 by Hiram S. Maxim. The explorer and colonialist Henry Morton Stanley (1841-1904) was extremely enthusiastic about Maxim's machine gun, and during a visit to the inventor he tried firing it, demonstrating that it really could fire 600 rounds per minute. Stanley commented that the machine gun would be "a valuable tool in helping civilization to overcome barbarism".*

New techniques in weaponry were introduced, and a naval armaments race began among the major industrial powers. The English found that their old navy was obsolete, and they had to rebuild. Thus, the period of colonial expansion between 1880 and 1914 was filled with tensions, as the industrial powers raced to arm themselves in competition with each other, and raced to seize as much as possible of the rest of the world. Industrial and colonial rivalry contributed to the outbreak of the First World War, to which the Second World War can be seen as a sequel.

4.2 Hobson's explanation

The English economist John Atkinson Hobson (1858-1940) offered a famous explanation for the colonial era in his book *Imperialism: A study* (1902). Hobson graduated from Lincoln College, Oxford, and later taught classics

Fig. 4.3 *A French cartoon from the 1890's showing England, Germany, Russia, France and Japan slicing up the pie of China.*

and English literature at schools in Faversham and Exeter. In 1887, he joined the Fabian Society and, during the last decade of the 19th century, he wrote several influential books: *Problems of Poverty*, (1891); *Evolution of Modern Capitalism*, (1894); *Problem of the Unemployed*, (1896); and *John Ruskin: Social Reformer*, (1898).

Hobson agreed with Ruskin's belief that economics should not be exclusively concerned with money matters but ought to contain ethical and humanitarian values as well, and he advocated the formation of cooperative labor guilds where human contacts would make work more pleasurable and rewarding.

The editor of the Manchester Guardian recruited John Hobson as a correspondent to cover the Second Boer War. His experiences in Africa as well as in England convinced Hobson that the war was being fought for economic reasons. In his book, *Imperialism*, published in 1902, Hobson analyzed the economic motivations behind the colonial era.

According to Hobson, the basic problem is an excessively unequal distribution of incomes in industrial countries like England. The result of this unequal distribution is that neither the rich nor the poor are in a position to buy back the total output of the highly industrialized nations. The poor cannot consume enough because their incomes are inadequate. Meanwhile the rich, who have enough money, are very few in number, and each of them has only finite needs. Therefore the rich cannot consume enough either, and they tend to save their excess money. The total effect is that the society is producing more than it can consume.

In this situation, Adam Smith would have proposed a simple solution: The rich (Smith would say) ought to reinvest their excess income in new factories. But, as Hobson pointed out, this would only aggravate the situation. If society is already unable to buy back its output, the new factories would only make matters worse by increasing production.

This situation, Hobson pointed out, provides a powerful economic motivation for imperialism. The excess output of industries can be sold to colonial peoples, and the excess savings of the rich can be invested abroad. This was in fact what was happening on a very large scale at the end of the 19th century. However, having personally witnessed the Second Boer War, Hobson believed imperialism to be immoral, since it entailed great suffering both among the colonial peoples and among the poor in the highly industrialized countries. The cure that Hobson recommended was a more equal distribution of incomes in the manufacturing nations.

Hobson was very popular as a lecturer and writer, but his ideas were too unorthodox to be accepted by the established economists of the time. His theory was, however, enthusiastically adopted by by V.I. Lenin, and Hobson's economic analysis of imperialism became a central part of Marxist-Leninist doctrine. This gave Hobson's ideas wide circulation, but in a political context that the mild mannered English economist would hardly have endorsed. Hobson's political opinions were in fact close to those of Ruskin and the Fabians, who believed in gradual progress rather than violent revolution.

4.3 The neocolonial era?

For a long time, Britain held its position as the leading industrial and colonial power, but from 1890 onwards its dominance was challenged by Germany, the United States, Belgium, France, Italy, Russia and Japan. Rivalry between these industrial powers, competing with each other for colonies, natural resources, markets, and military power, contributed to the start of World War I. At the end of "the Great War", the League of Nations assigned "protectorates" to the victors. These "protectorates" were, in fact, colonies with a new name, although in principle protectorates were supposed to be temporary.

The Second World War was terrible enough to make world leaders resolve to end the institution of war once and for all, and the United Nations was set up for this purpose. Despite the flaws and weaknesses of the UN Charter, the organization was successful in formally ending the era of colonialism. One must say "formally ending" rather than "ending", because colonialism persisted in a new guise: During the classical era of colonialism, there was direct political power, with Viceroys and Governors General acting as formal rulers of colonies. During the decades following the Second World War, almost all colonies were granted formal independence, but nevertheless the influence of the industrialized nations was strongly felt in the developing world. Direct political power was replaced by indirect methods.

Suggestions for further reading

(1) E.J. Hobsbawn, *The Age of Empire, 1875-1914*, Vintage Books, (1989).
(2) L. James, *The Rise and Fall of the British Empire*, St Martin's Press, (1997).
(3) N. Ferguson, *Empire: The Rise and Demise of the British World Order and the Lessons for Global Power*, Basic Books, (2003).
(4) S. Schama, *The Fate of Empire, 1776-2000*, Miramax, (2002).
(5) A.P. Thorton, *The Imperial Idea and Its Enemies: A Study in British Power*, Palgrave Macmillan, (1985).
(6) H. Mejcher, *Imperial Quest for Oil: Iraq, 1910-1928*, Ithaca Books, London, (1976).
(7) P. Sluglett, *Britain in Iraq, 1914-1932*, Ithaca Press, London, (1976).
(8) D.E. Omissi, *British Air Power and Colonial Control in Iraq, 1920-1925*, Manchester University Press, Manchester, (1990).
(9) V.G. Kiernan, *Colonial Empires and Armies, 1815-1960*, Sutton, Stroud, (1998).
(10) R. Solh, *Britain's 2 Wars With Iraq*, Ithaca Press, Reading, (1996).

(11) D. Hiro, *The Longest War: The Iran-Iraq Military Conflict*, Routledge, New York, (1991).
(12) T.E. Lawrence, *A Report on Mesopotamia by T.E. Lawrence*, Sunday Times, August 22, (1920).
(13) D. Fromkin, *A Peace to End All Peace: The Fall of the Ottoman Empire and the Creation of the Modern Middle East*, Owl Books, (2001).
(14) T. Rajamoorthy, *Deceit and Duplicity: Some Reflections on Western Intervention in Iraq*, Third World Resurgence, March-April, (2003).
(15) P. Knightley and C. Simpson, *The Secret Lives of Lawrence of Arabia*, Nelson, London, (1969).
(16) G. Lenczowski, *The Middle East in World Affairs*, Cornell University Press, (1962).
(17) Y. Nakash, *The Shi'is of Iraq*, Princeton University Press, (1994).
(18) G. Kolko, *Another Century of War*, New Press, (2002).
(19) D. Yergin, *The Prize*, Simon and Schuster, New York, (1991).
(20) A. Sampson, *The Seven Sisters: The Great Oil Companies of the World and How They Were Made*, Hodder and Staughton, London, (1988).
(21) J.D. Rockefeller, *Random Reminiscences of Men and Events*, Doubleday, New York, (1909).
(22) M.B. Stoff, *Oil, War and American Security: The Search for a National Policy on Oil, 1941-1947*, Yale University Press, New Haven, (1980).
(23) W.D. Muscable, *George F. Kennan and the Making of American Foreign Policy*, Princeton University Press, Princeton, (1992).
(24) J. Stork, *Middle East Oil and the Energy Crisis*, Monthly Review, New York, (1976).
(25) F. Benn, *Oil Diplomacy in the Twentieth Century*, St. Martin's Press, New York, (1986).
(26) K. Roosevelt, *Countercoup: The Struggle for the Control of Iran*, McGraw-Hill, New York, (1979).
(27) E. Abrahamian, *Iran Between Two Revolutions*, Princeton University Press, Princeton, (1982).
(28) J.M. Blair, *The Control of Oil*, Random House, New York, (1976).
(29) M.T. Klare, *Resource Wars: The New Landscape of Global Conflict*, Owl Books reprint edition, New York, (2002).
(30) M. Muffti, *Sovereign Creations: Pan-Arabism and Political Order in Syria and Iraq*, Cornell University Press, (1996).
(31) C. Clover, *Lessons of the 1920 Revolt Lost on Bremer*, Financial Times, November 17, (2003).
(32) J. Kifner, *Britain Tried First. Iraq Was No Picnic Then*, New York Times, July 20, (2003).

(33) D. Omissi, *Baghdad and British Bombers*, Guardian, January 19, (1991).
(34) D. Vernet, *Postmodern Imperialism*, Le Monde, April 24, (2003).
(35) J. Buchan, *Miss Bell's Lines in the Sand*, Guardian, March 12, (2003).
(36) C. Tripp, *Iraq: The Imperial Precedent*, Le Monde Diplomatique, January, (2003).
(37) C. Johnson, *The Sorrows of Empire: Militarism, Secrecy, and the End of the Republic*, Henry Hold and Company, New York, (2004).
(38) C. Johnson, *Blowback: The Costs and Consequences of American Empire*, Henry Hold and Company, New York, (2000).
(39) M. Parenti, *Against Empire: The Brutal Realities of U.S. Global Domination*, City Lights Books, 261 Columbus Avenue, San Francisco, CA94133, (1995).
(40) E. Ahmad, *Confronting Empire*, South End Press, (2000).
(41) S.R. Shalom, *Imperial Alibis*, South End Press, (1993).
(42) J.K. Galbraith, *The Unbearable Costs of Empire*, American Prospect magazine, November, (2002).
(43) G. Monbiot, *The Logic of Empire*, The Guardian, August 6, (2002), World Press Review, October, (2002).
(44) J. Wilson, *Republic or Empire?*, The Nation magazine, March 3, (2003).
(45) P. Cain and T. Hopkins, *British Imperialism, 1688-2000*, Longman, (2000).
(46) M.T. Klare, *Geopolitics Reborn: The Global Struggle Over Oil and Gas Pipelines*, Current History, December issue, 428-33, (2004).

Chapter 5

MAINSTREAM INDUSTRIALISM

5.1 Trade unions and minimum wage laws

Robert Owen and social reform

During the early phases of the Industrial Revolution in England, the workers suffered greatly. Enormous fortunes were made by mill and mine owners, while workers, including young children, were paid starvation wages for cruelly long working days. However, trade unions, child labor laws, and the gradual acceptance of birth control finally produced a more even distribution of the benefits of industrialization.

One of the most interesting pioneers of these social reforms was Robert Owen (1771-1858), who is generally considered to have been the father of the Cooperative Movement. Although in his later years not all of his projects developed as he wished, his life started as an amazing success story. Owen's life is not only fascinating in itself; it also illustrates some of the reforms that occurred between 1815 and 1850.

Robert Owen was born in Wales, the youngest son of a family of iron-mongers and saddle-makers. He was a very intelligent boy, and did well at school, but at the age of 9, he was apprenticed to a draper, at first in Wales. Later, at the age of 11, he was moved to London, where he was obliged to work eighteen hours a day, six days a week, with only short pauses for meals. Understandably, Robert Owen found this intolerable, and he moved again, this time to Manchester, where he again worked for a draper.

While in Manchester, Robert Owen became interested in the machines that were beginning to be used for spinning and weaving. He borrowed a hundred pounds from his brother, and entered (as a partner) a small business that made these machines. After two years of moderate success as a small-scale industrialist, Owen saw the newspaper advertisement of a position for manager of a large spinning mill, owned by a Mr. Drinkwater.

"I put on my hat", Owen wrote later, "and proceeded straight to Mr.

Drinkwater's counting house. 'How old are you?' 'Twenty this May', was my reply. 'How often do you get drunk in the week?'... 'I was never', I said, 'drunk in my life.' blushing scarlet at this unexpected question. 'What salary do you ask?' 'Three hundred a year', was my reply. 'What?', Mr. Drinkwater said with some surprise, repeating the words, 'Three hundred pounds! I have had this morning I know not how many seeking the situation and I do not think that all of their askings would amount to what you require.' 'I cannot be governed by what others seek', said I, 'and I cannot take less.'

Apparently impressed by Robert Owen's success as a small-scale industrialist, and perhaps also impressed by his courage, Mr. Drinkwater hired him. Thus, at the age of 19, Owen became the manager of a large factory. Mr. Drinkwater had no cause to regret his decision, since his new manager quickly became the boy wonder of Manchester's textile community. Within six months, Drinkwater offered Owen a quarter interest in his business.

After several highly successful years in his new job, Robert Owen heard of several mills that were for sale in the village of New Lanark, near to Glasgow. The owner, Mr. Dale, happened to be the father of the girl with whom Robert Owen had fallen in love. Instead of directly asking Dale for permission to marry his daughter, Owen (together with some business partners) first purchased the mills, after which he won the hand of the daughter.

Ownership of the New Lanark mills gave Robert Owen the chance to put into practice the ideas of social reform that he had been developing throughout his life. Instead of driving his workers by threats of punishment, and instead of subjecting them to cruelly long working hours (such as he himself had experienced as a draper's apprentice in London), Owen made the life of his workers at New Lanark as pleasant as he possibly could. He established a creche for the infants of working mothers, free medical care, concerts, dancing, music-making, and comprehensive education, including evening classes. Instead of the usual squalid one-room houses for workers, neat two-room houses were built. Garbage was collected regularly instead of being thrown into the street. New Lanark also featured pleasant landscaped areas.

Instead of leading to bankruptcy, as many of his friends predicted, Robert Owen's reforms led to economic success. Owen's belief that a better environment would lead to better work was vindicated. The village, with its model houses, schools and mills, became internationally famous as a demonstration that industrialism need not involve oppression of the workers. Crowds of visitors made the journey over narrow roads from Glasgow to learn from New Lanark and its visionary proprietor. Among the twenty thousand visitors who signed the guest-book between 1815 and 1825 were

Fig. 5.1 *New Lanark World Heritage village in Scotland. A view of the school.*

the Grand Duke Nicholas of Russia (who later became Czar Nicholas I), and Princes John and Maximilian of Austria.

Robert Owen's ideas of social reform can be seen in the following extract from an "Address to the Inhabitants of New Lanark", which he presented on New Year's Day, 1616: "What ideas individuals may attach to the term 'Millennium' I know not; but I know that society may be formed so as to exist without crime, without poverty, with health greatly improved, with little, if any, misery. and with intelligence and happiness increased a hundredfold; and no obstacle whatsoever intervenes at this moment except ignorance to prevent such a state of society from becoming universal."

Robert Owen believed that these principles could be applied not only in New Lanark but also in the wider world. He was soon given a chance to express this belief. During the years from 1816 to 1820, apart from a single year, business conditions in England were very bad, perhaps as a result of the Napoleonic Wars, which had just ended. Pauperism and social unrest were widespread, and threatened to erupt into violence. A committee to deal with the crisis was formed under the leadership of the Dukes of Kent and York.

Because of Owen's reputation, he was asked for his opinion, but the committee was hardly expecting the answer that they received from him. Robert Owen handed the two Dukes and the other committee members a detailed plan for getting rid of pauperism by making paupers productive. They were to be settled in self-governing Villages of Cooperation, each with between 800 and 1,200 inhabitants. Each family was to have a private apartment, but there were to be common sitting rooms, reading rooms and kitchens. Near to the houses, there were to be gardens tended by the children, and farther out, fields to be cultivated by the adults. Still farther from the houses, there was to be a small factory.

Owen's idea for governmentally-planned paupers' collectives was at first rejected out of hand. The early 19th century was, after all, a period of unbridled *laissez-faire* economics. Owen then bombarded the Parliament with pamphlets advocating his scheme. Finally a committee was formed to try to raise the money to establish one Village of Cooperation as an experiment; but the money was never raised.

Unwilling to accept defeat, Robert Owen sold his interest in New Lanark and sailed for America, where he believed that his social experiment would have a better chance of success. He bought the town of Harmonie and 30,000 acres of land on the banks of the Wabash River in Indiana. There he established a Village of Cooperation which he named "New Harmony". He dedicated it on the 4th of July, 1826. It remained a collective for only two years, after which individualism reasserted itself. Owen's four sons and one of his daughters made their homes in New Harmony, and it also became the home of numerous scientists, writers and artists.

Owen's son, Robert Dale Owen, became a member of the U.S. House of Representatives, where he introduced the bill establishing the Smithsonian Institution. In 1862 he wrote an eloquent letter to Abraham Lincoln urging emancipation of the slaves. Three days later, probably influenced by Owen's letter, Lincoln read the Emancipation Proclamation to his cabinet. Another son, Richard Owen, served as President of the University of Indiana, and was later elected as the first President of Purdue University.

When Robert Owen returned to England shortly after dedicating New Harmony, he found that he had become a hero of the working classes. They had read his writings avidly, and had begun to establish cooperatives, following his principles. There were both producer's cooperatives and consumer's cooperatives. In England, the producer's cooperatives failed, but in Denmark they succeeded[1].

One of the early consumer's cooperatives in England was called the Rochdale Society of Equitable Pioneers. It was founded by 28 weavers and

[1]The success of Danish agricultural producer's cooperatives was helped by the People's High School movement, founded by N.F.S. Grundvig (1783-1872).

other artisans, who were being forced into poverty by mechanization. They opened a small cooperative store selling butter, sugar, flour, oatmeal and candles. After a few months, they also included tobacco and tea. From this small beginning, the Cooperative Movement grew, finally becoming one of the main pillars of the British Labour Party.

Fig. 5.2 *Robert Owen, (1771-1858), founder of the Cooperative Movement.*

Robert Owen's attention now turned from cooperatives to the embryonic trade union movement, which was struggling to establish itself in the face of fierce governmental opposition. He assembled the leaders of the working class movement and proposed the formation of the "Grand National Moral Union of Productive and Useful Classes". The name was soon shortened to "The Grand National Consolidated Trades Union" or simply the "Grand National".

Owen's Grand National was launched in 1833, and its membership quickly grew to half a million. It was the forerunner of modern nationwide trade unions, but it lasted only two years. Factory-owners saw the Grand National as a threat, and they persuaded the government to prosecute it under anti-union laws. Meanwhile, internal conflicts helped to destroy the Grand National. Owen was accused of atheism by the working class leaders, and he accused them of fermenting class hatred.

Robert Owen's influence helped to give raw *laissez faire* capitalism a more human face, and helped to spread the benefits of industrialization more widely. Through the work of other reformers like Owen, local trade unions succeeded, both in England and elsewhere; and in the end, successful national unions were finally established. The worst features of the early Industrial Revolution were moderated by the growth of the trade union movement, by child labor laws, by birth control and by minimum wage laws.

Rusting of the Iron Law

David Ricardo's Iron Law of Wages maintained that workers must necessarily live at the starvation level: Their wages are determined by the law of supply and demand, Ricardo said. If the wages should increase above the starvation level, more workers' children would survive, the supply of workers would increase, and the wages would fall again. This gloomy pronouncement was enthusiastically endorsed by members of the early 19th century Establishment, since it absolved them from responsibility for the miseries of the poor. However, the passage of time demonstrated that the Iron Law of Wages held only under the assumption of an economy totally free from governmental intervention.

Both the growth of the political power of industrial workers, and the gradual acceptance of birth control were important in eroding Ricardo's Iron Law. Birth control is especially important in countering the argument used to justify child labor under harsh conditions. The argument (still used in many parts of the world) is that child labor is necessary in order to save the children from starvation, while the harsh conditions are needed because if a business provided working conditions better than its competitors, it would go out of business. However, with a stable population and appropriate social legislation prohibiting both child labor and harsh working conditions, the Iron Law argument fails.

5.2 Rising standards of living

Since the year 1000, world population has risen 22-fold, global per capita Gross Domestic Product 13-fold, and world GDP nearly 300-fold. These data come from Angus Maddison's recent book, *World Population, GDP and Per Capita GDP, 1-2003*. More detailed data, from a report that Prof. Maddison presented to the British House of Lords, are shown in Tables 5.1 and 5.2.

During the period between 1820 and 2001, the average years of education per person employed increased from 2.00 years to 15.45 years in the United Kingdom, from 1.75 years to 20.21 years in the United States, and from 1.50 years to 16.61 years in Japan. This increased education in the highly industrialized countries was necessary because of the complexity of modern machines and modern life.

Today, most citizens of the industrialized countries have lives of greatly-increased pleasure and freedom compared with the lives of their great-grandparents. Furthermore, their lives are also remarkably easy and pleasant compared with the remainder of the world. In later chapters we will try to discuss to what extent this privileged life-style is sustainable.

5.3 Robber barons and philanthropists

"Hain't I got the power?"

We can experience some of the flavor of early American industrial growth by looking at the life of Cornelius Vanderbilt (1794-1877). In those days, the United States was a place where a man with luck, intelligence and energy, could start with nothing and become a multimillionaire. That is exactly what Vanderbilt did.

Vanderbilt was born into a poor New York family. He quit school at 11 to help his father, and later remarked, "If I had learned education, I wouldn't have had time to learn anything else." At 16 he started his first business, using $100 borrowed from his mother - a small ferry boat between New York and Staten Island, charging 18 cents per trip. The business succeeded because of the fair price that he charged and because of his prodigious work. Within a year, he was able to give his mother $1,000 in return for her loan.

Table 5.1: GDP per capita (1990 int. $). Data from Maddison.

	1900	1950	1990	2001
W. Europe	2,893	4,579	15,966	19,256
USA	4,091	9,561	23,201	27,948
Ca.,Au.,NZ	3,435	7,424	17,902	21,718
Japan	1,180	1,921	18,789	20,683
E. Europe	1,438	2,111	5,450	6,207
fUSSR	1,237	2,841	6,878	4,626
L. America	1,109	2,506	5,053	5,811
China	545	439	1,858	3,583
India	599	619	1,309	1,957
Other Asia	802	919	3,084	3,997
Africa	601	894	1,444	1,489
World	1,262	2,111	5,157	6,049

Table 5.2: Gross stock of machinery and equipment per capita (1990 $). Data from Maddison. These figures are a measure of the degree of industrialization of the countries shown. Similar increases occurred in the gross stock of non-residential structures per capita. For example, in the USA the value of these structures increased from $1,094 (1990 $) in 1820 to $36,330 in 2001. In Japan there was a dramatic increase during the 20th century, from $852 per capita in 1913 to $57,415 in 2001.

	UK	USA	Japan
1820	92	87	na
1870	334	489	94
1913	878	2,749	329
1950	2,122	6,110	1,381
1973	6,203	10,762	6,431
2001	16,082	30,600	32,929

During the War of 1812, Vanderbilt had a government contract to sail supplies to forts in the New York area. He was by then operating a small fleet of sailing schooners, and as a consequence he received the nickname, "Commodore".

Cornelius Vanderbilt then became interested in steamships, but Robert Fulton and Robert Livingston had been granted a 30-year monopoly on the steamboat trade. This did not stop Vanderbilt. He started a competing steam line, and his boat evaded capture. Finally a Supreme Court decision broke the Fulton-Livingston monopoly. By the 1840's, Vanderbilt was op-

Fig. 5.3 *Cornelius "Commodore" Vanderbilt.*

erating about 100 steamships, and his business had the most employees of any in the United States.

Turning his attention to railways, Vanderbilt bought several lines, including the New York and Harlem Railroad, the Hudson River Railroad, and the New York Central Railroad. He extended his lines as far as Chicago, and attempted to acquire the Erie Railroad. This brought him into conflict with the unscrupulous financier Jim Fisk. Vanderbilt's methods were equally rough, so it was a fight with no holds barred. (Cornelius Vanderbilt once remarked, "What do I care about the law? Hain't I got the power?")

At the time of his death, Cornelius Vanderbilt was one of the richest men in the United States, with a fortune of over $100,000,000. He left most of this amount to his son William [2], but gave one million to Central University, which then became Vanderbilt University.

[2]William Vanderbilt is best remembered for his remark, "The public be damned!"

Carnegie's philanthropies

We can contrast Vanderbilt's relatively small interest in philanthropy with Andrew Carnegie's large-scale efforts for public improvement. Like Vanderbilt, Andrew Carnegie (1835-1919) was a self-made multimillionaire, but after making a fortune in oil wells, steel, iron ore and railways, he gave almost all of his money away. Early in his career, he wrote:

Fig. 5.4 *Andrew Carnegie circa 1878.*

"I propose to take an income no greater than $50,000 per annum! Beyond this I need never earn, make no effort to increase my fortune, but spend the surplus each year for benevolent purposes! Let us cast aside business forever, except for others. Let us settle in Oxford, and I shall get a thorough education, making the acquaintance of literary men... To continue much longer overwhelmed by business cares and with most of my thoughts wholly upon the way to make more money in the shortest time, must degrade me beyond hope of permanent recovery."

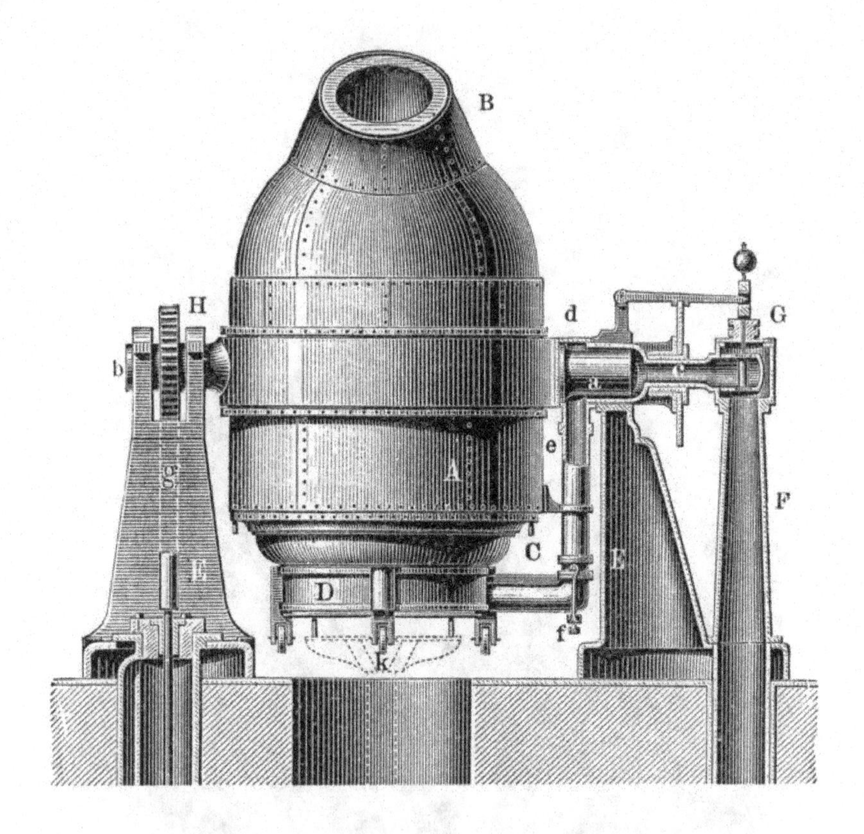

Fig. 5.5 *A Bessemer converter, used in making steel.*

When he sold his share of the United States Steel Corporation in 1901, Andrew Carnegie became one of the wealthiest men in the world. He devoted the remainder of his life to educational projects and to philanthropy. He established a large number of public libraries, not only in the United Kingdom and in the United States, but also in Canada, Ireland, Australia,

New Zealand, the West Indies and Fiji. In all, Carnegie established 3,000 libraries. In addition, he founded the Carnegie Institution in Washington D.C. and the Carnegie Institute of Technology in Pittsburgh, which later became the Carnegie Mellon University.

In Scotland, his birthplace, where he lived for part of each year, Andrew Carnegie established a trust to assist in university education. In recognition of this generous gift (and perhaps also in recognition of his authorship of a number of books and articles), Andrew Carnegie was elected Lord Rector of the University of St. Andrews. Carnegie also gave a large amount of money to Booker T. Washington's Tuskegee Institute. He established generous pension funds for his former employees, and also for American university professors. As if all this were not enough, he paid for the construction of 7,000 church organs, contributed to the erection of the Peace Palace at the Hague, and established the Carnegie Foundation, which continues to perform good works, especially in the field of education.

In the lives of Cornelius Vanderbilt and Andrew Carnegie we see exemplified some of the features of the age in which they lived, when ruthless business behavior was often balanced by splendid acts of public generosity.

5.4 The conflict between capitalism and communism

The Russian Revolution

Industrialism in Russia started more slowly than in Europe and the United States. The emancipation of the surfs in 1861 by Czar Alexander II was incompletely carried out in practice, and Russia remained, to a large extent, in the grip of feudal absolutism. Dissatisfaction with the slowness of reform led to a series of protests. On January 22, 1905 (Bloody Sunday), a group of marchers in St. Petersburg brought a petition to the Winter Palace, hoping to hand it to the Czar. The petition called for improvement of working conditions, democratic elections, and establishment of a constituent assembly. Without the Tzar's knowledge (he was not in St. Petersburg at the time), government troops fired on the marchers, and about 1,000 were killed.

Between 1905 and 1908, strikes and peasant disorders occurred throughout Russia. The revolt of sailors on the Battleship Potemkin, upon which Sergei Eisenstein based his famous film, also occurred at this time.

In 1914, Russia entered the First World War on the side of England and France. The Russian troops were badly supplied, and suffered heavy casualties. With the soldiers absent from their fields, some of the larger Russian cities were threatened with famine.

By February, 1917, dissatisfaction had reached such a level that a total general strike occurred in St Petersburg. Instead of putting down this strike, weary Russian soldiers supported it, handing over their weapons to members of the angry crowds. These events lead to the abdication of Czar Nicholas II, who handed over power to the Kerensky Provisional Government. In October, 1917, the Bolsheviks under the leadership of V.I. Lenin, gained power. Lenin had been exile in Switzerland before returning to lead the October Revolution, and he was a follower of the communist economist, Karl Marx.

The First Red Scare; McCarthyism; the Cold War

There were two distinct periods of violent anti-communism in the United States. The "First Red Scare" occurred between 1917 and 1920, while "McCarthyism" began in the late 1940's and lasted until the late 1950's.

The "First Red Scare" was largely inspired by the fear that the 1917 revolution in Russia would spread to the United States. In 1919, a bomb plot was uncovered; bombs were to be sent through the post to 36 prominent Americans, including John Pierpont Morgan, John D. Rockefeller and Supreme Court Justice Oliver Wendell Holmes. The year 1919 was also characterized by hundreds of strikes throughout the United States. Newspapers described the strikes as communist plots, and the FBI arrested several thousand suspected agitators.

The McCarthy era occurred after World War II. The United States emerged from the war as the only major industrial power whose infrastructure had not been destroyed by the war. Thus the US found itself thrust somewhat reluctantly and nervously into a position of global leadership. Meanwhile a communist revolution had occurred in China, and this added to US nervousness, as did the Soviet development of nuclear weapons.

Senator Joseph McCarthy (1908-1957), and the House Un-American Activities Committee lead an aggressive populistic hunt for communists and "communist sympathizers". About 500 Hollywood actors, actresses and screenwriters were blacklisted.

The end of World War II also marked the start of the "Cold War" between capitalist and communist countries. The most dangerous feature of the Cold War was a nuclear arms race that resulted in a truly insane number of nuclear weapons. At the height of this arms race there were over 50,000 nuclear weapons in the world, with a total explosive power roughly a million times greater than the bomb that destroyed Hiroshima in 1945. Put another way, the bombs had an explosive power equivalent to 4 tons of TNT for every person on the planet. The world came close to thermonuclear war on several occasions, for example during the Cuban

Missile Crisis of October, 1962. Although the Cold War has now ended, about 27,000 nuclear weapons still exist, many of them on hair-trigger alert. Because of the dangers of accidental nuclear war, nuclear proliferation and nuclear terrorism, these weapons continue to cast a very dark shadow over the future of humankind.

Capitalism triumphant

After the fall of the Berlin Wall in 1989, and the dissolution of the Soviet Union in 1991, capitalism spread to much of what had been the communist block of nations. Even China, although remaining officially a communist state, adopted capitalist methods on an experimental basis. Adam Smith, the prophet of the free market and of economic growth, was triumphant. We should notice that despite their differences regarding ownership of the means of production, capitalists and communists are united in their admiration of economic growth.

5.5 Globalization

In Chapter 3, we mentioned the exploitation of factory workers during the early phases of the Industrial Revolution. In the present chapter, we discussed how the growth of trade unions, the enactment of minimum wage laws, and laws preventing child labor, together with the gradual acceptance of birth control, led to a more widely-distributed prosperity, where workers shared the benefits of industrialization.

Today, economic globalization aims at increased trade throughout the world. At first sight, this seems to be a benefit. However, laws preventing the exploitation of labor are not universal. Industrialization of some developing countries has repeated the the abuses that were discussed in Chapter 3, so that in many developing countries, it involves slave-like working conditions. Workers in the developed countries can find themselves competing with grossly underpaid labor in developing nations. The cure, of course, is to demand universal laws protecting workers from exploitation. Such laws must be a precondition for free trade.

5.6 Say's law

Suburbia

The private automobile is the flagship of industrialism. In 2002, there were more than half a billion automobiles in the world. Of these, 140 million were in the United States (roughly one for every two people).

Reliance on private automobiles for transportation has affected the geography of cities, producing vast highway systems, urban sprawl and suburban life. For example, the Los Angeles metropolitan area spreads over 4,850 square miles (12,400 km^2). Because of the availability of inexpensive motor fuel, public transportation is almost non-existent in Los Angeles. It is not uncommon for a citizen of the city to drive several hundred kilometers during a normal day. Many other cities in the world have a similar dependence on private automobiles.

A recent Canadian documentary film, *The End of Suburbia*, explores the history and probable future of cities built around the availability of inexpensive gasoline. The subtitle of the film is *Oil Depletion and the Collapse of the American Dream.*

Keeping up appearances

Of course, if we live in suburbia, we have to keep up with the neighbors. This is hard to do, because the neighbors keep getting new things - bigger automobiles, motorboats, swimming pools, and so on. Not only must we keep up with our actual neighbors, we must also compete with the glamorous lives that we see in films and television.

According to Say's Law, and according to advertisers and economists, human desires have no upper limit; there is no limit to growth. Television advertising and billboards constantly tell us that to be happy, or even respectable, we need to buy more. Thus mainstream industrial culture thunders ahead, worshiping power, material goods, wealth, growth and progress. There is, however, a counterculture, which we will look at in the next chapter.

Suggestions for further reading

(1) R. Owen, *A New View of Society, or, Essays on the Formation of the Human Character Preparatory for the Development of a Plan for Gradually Ameliorating the Condition of Mankind*, Longman, London, (1916).
(2) R. Owen, *The Life of Robert Owen, by Himself*, ed. M. Beer, Knopf, New York, (1920).
(3) R. Podmore, *Robert Owen, A Biography*, Allan and Unwin, (1906).
(4) G.D.H. Cole, *Life of Robert Owen*, Macmillan, (1930).
(5) J. Butt, ed., *Robert Owen: Prince of Cotton Spinners*, David and Charles, (1971).
(6) G. Claeys, ed., *A New View of Society and other writings by Robert Owen*, Penguin Classics, (1991).

(7) G. Claeys, ed., *Selected Works of Robert Owen in 4 volumes*, Pickering, (1993).
(8) R. Sobel, *The Big Board: A History of the New York Stock Market*, Beard Books, (2000).
(9) A. Kohn, *No Contest - The Case Against Competition*, Houghton Mifflin Co., (1986).
(10) A.T. Vanderbilt, *Fortune's Children: The Fall of the House of Vanderbilt*, William Morrow, New York, (1989).
(11) D. Nasaw, *Andrew Carnegie*, Penguin Press, New York, (2006).
(12) J.R.T. Hughes, *The Vital Few: American Economic Progress and its Protagonists*, Houghton and Mifflin, Boston, (1965).
(13) H. Livesay, *Andrew Carnegie and the Rise of Big Business*, Houghton Mifflin, Boston, (1975).
(14) G. Wright, *The Origins of American Industrial Success, 1879-1940*, American Economic Review, **80**, 651-668, (1990).
(15) A. Carnegie, *Autobiography of Andrew Carnegie*, Houghton Mifflin, Boston, (1920).
(16) A. Carnegie, *Triumphant Democracy, or, Fifty Year's March of the Republic*, Scribners, New York, (1886).
(17) A. Maddison, *The World Economy: A Millenial Perspective*, Overseas Press, New Delhi, (2003).
(18) M. Steinberg, *Voices of Revolution, 1917*, Yale University Press, (2001).
(19) R. Malone, *Analysing the Russian Revolution*, Cambridge University Press, Melbourne, (2004).
(20) O. Figes, *A People's Tragedy: The Russian Revolution, 1891-1924*, ISBN 0-14-024364-X.
(21) M.B. Levin, *Political Hysteria in America: The Democratic Capacity for Repression*, Basic Books, (1971).
(22) J.E. Haynes, *Red Scare of Red Menace? American Communism and Anti Communism in the Cold War Era*, Ivan R. Dee, (2000).
(23) A. Fried, *McCarthyism, The Great American Red Scare: A Documentary History*, Oxford University Press, (1997).
(24) T. Morgan, *Reds: McCarthyism in Twentieth-Century America*, Random House, (2004).
(25) E. Schrecker, *Many Are the Crimes: McCarthyism in America*, Little, Brown, (1998).
(26) R. Fishman, *Bourgois Utopias: The Rise and Fall of Suburbia*, Basic Books, (1987).
(27) R. Fishman, *America's New City: Megalopolis Unbounded*, Wilson Quarterly, **14**, 24-45, (1990).

(28) J. Borchert, *Residential City Suburbs: The Emergence of a New Suburban Type, 1880-1930*, Journal of Urban History, **22**, 283-307, (1996).
(29) K.A. Daniellsen et al., *Retracting Suburbia: Smart Growth and the Future of Housing*, Housing Policy Debate, **10**, 513-540, (1999).
(30) J. Garreau, *Edge City: Life on the New Frontier*, Doubleday, New York, (1991).
(31) R.E. Lang, *Edgeless Cities: Exploring the Elusive Metropolis*, Brookings Institution Press, Washington D.C., (2002).
(32) S.B. Warner, *Streetcar Suburbs: The Process of Growth in Boston 1870-1890*, Cambridge Mass., (1962).
(33) K.M. Kruse and T.J. Sugrue, eds., *The New Suburban History*, University of Chicago Press, (2006).
(34) B. Kelly, *Expanding the American Dream: Building and Rebuilding Levittown*, State University of Albany Press, Albany NY, (1993).
(35) J.E. Stiglitz, *Globalization and its Discontents*, W.W. Norton, New York, (2002).
(36) J.E. Stiglitz, *Making Globalization Work*, W.W. Norton, New York, (2006).
(37) M. Steger, *Globalization: A Very Short Introduction*, Oxford University Press, (2003).
(38) A. MacGillivray, *A Brief History of Globalization: The Untold Story of our Incredibly Shrinking Plannet*, Carroll and Graf, (2006).
(39) T.L. Friedman, *The World is Flat*, Farrar, Straus and Giroux, (2006).
(40) J. Pilger, *The New Rulers of the World*, Verso Books, (2003).
(41) R.J. Barrow, *Determinants of Economic Growth: A Cross-Country Empirical Study*, MIT Press, Cambridge MA, (1997).
(42) D.K. Foley, *Growth and Distribution*, Harvard University Press, (1999).
(43) C.I. Jones, *Introduction to Economic Growth, 2nd ed.*, W.W. Norton, (2002).
(44) S. Vaclav, *China's Environmental Crisis: An Inquery into the Limits of National Development*, M.E. Sharpe, Armonk, (1992).
(45) J.D. Hammond and C.H. Claire, eds., *Making Chicago Price Theory: Friedman-Stigler Correspondence 1945-1957*, , Routledge, (2006).

Chapter 6

VEBLEN, GANDHI AND THE GREENS

6.1 Veblen; economics as anthropology

The phrase "conspicuous consumption" was invented by the Norwegian-American economist Thorstein Veblen (1857-1929) in order to describe the way in which our society uses economic waste as a symbol of social status. In *The Theory of the Leisure Class*, first published in 1899, Veblen pointed out that it is wrong to believe that human economic behavior is rational, or that it can be understood in terms of classical economic theory. To understand it, Veblen maintained, one might preferably make use of insights gained from anthropology, psychology, sociology, and history.

Thorstein Veblen was born into a large Norwegian immigrant family living on a farm in Wisconsin. His first language was Norwegian, and in fact he did not learn English well until he was in his teens. He was a strange boy, precociously addicted to reading, but negligent about doing his chores on the farm. His family recognized that he was unusually intelligent and decided to send him to Carlton College, where he obtained a B.A. in 1880. Later he did graduate work at Johns Hopkins University and finally obtained a Ph.D. from Yale in 1884.

Despite the Ph.D., he failed to obtain an academic position. His iconoclastic views and non-conformist attitudes undoubtedly contributed to this joblessness. Returning to the family farm, Thorstein Veblen continued his voracious reading and his neglect of farm duties for six years. As one of his brothers wrote, "He was lucky enough to come out of a race and family who made family loyalty a religion... He was the only loafer in a highly respectable community... He read and loafed, and the next day he loafed and read."

An interesting fact about this strange man is that, for some reason, women found him very attractive. In 1888, Thorstein Veblen married Ellen Rolfe, the niece of the president of Carlton College. His wife was to leave him many times, partly because of his many infidelities, and partly because of his aloofness and detachment. He was like a visitor from another planet.

Fig. 6.1 *Thorstein Veblen (1857-1929).*

In part, the marriage to Ellen was motivated by Veblen's search for a job. He hoped to obtain work as an economist for the Atchison, Topeka and Santa Fe Railway, of which her uncle was president. However, the railway was in financial difficulties, and it was taken over by bankers, after which the position disappeared.

Finally a family council was held on the Veblen farm, and it was decided that Thorstein should once again attempt to enter the academic world. In 1891, wearing corduroy trousers and a coonskin hat, he walked into the office of the conservative economist J.L. Laughlan and introduced himself. Although taken aback by Veblen's appearance, Laughlan began to talk with him, and he soon recognized Veblen's genius. A year later, when he moved to the University of Chicago, Laughlan brought Veblen with him at a salary of $520 per year.

At the University of Chicago, Veblen soon established a reputation both for eccentricity and for enormous erudition. His socks were held up by safety pins, but he was reputed to be fluent in twenty-six languages. He gained attention also by publishing a series of brilliant essays.

In 1899, Veblen "fluttered the dovecotes of the East" by publishing a book entitled *The Theory of the Leisure Class.* It was part economics, part anthropology, and part social satire. Nothing of the kind had ever been seen in the field of economics. Until that moment it had been universally assumed that human economic behavior is rational. Veblen's detached and surgically sharp intelligence exposed it as being very largely irrational.

According to Thorstein Veblen, ancient tribal instincts and attitudes motivate us today, just as they motivated our primitive ancestors. Veblen speaks of a predatory phase of primitive society where the strongest fighters were able to subjugate others. This primitive class structure was based on violence, and, according to Veblen, the attitudes associated with it persist today.

For example, Veblen noted that male members of the leisure class liked to go about with walking sticks. Why? Because, answers Veblen, it is "an advertisement that the bearer's hands are employed otherwise than in useful effort." Also, a walking stick is a weapon: "The handling of so tangible and primitive a means of offense is very comforting to anyone who is gifted with even a moderate share of ferocity".

Even in modern society, Veblen says, we have an admiration for those who succeed in obtaining power and money through predatory means, and this admiration makes honest and useful work seem degraded. "During the predatory culture", Veblen wrote, "labour comes to be associated in men's habits of thought with weakness and subjugation to a master. It is therefore a mark of inferiority, and therefore comes to be accounted to be unworthy of man in his best estate. By virtue of this tradition, labour is felt to be debasing, and this tradition has never died out. On the contrary, with the advance of social differentiation it has acquired the axiomatic force of ancient and unquestioned prescription."

"In order to gain and hold the esteem of men it is not sufficient merely to possess wealth or power. The wealth or power must be put in evidence, for esteem is awarded only on evidence. It is felt by all persons of refined taste that a spiritual contamination is inseparable from certain offices that are conventionally required of servants. Vulgar surroundings, mean (that is to say, inexpensive) habitations, and vulgarly productive occupations are unhesitatingly condemned and avoided. They are incompatible with life on a satisfactory spiritual plane - with 'high thinking'."

"...The performance of labour has been accepted as a conventional evidence of inferior force, therefore it comes by itself, by a mental shortcut, to be regarded as intrinsically base."

"The normal and characteristic occupations of the [leisure] class are... government, war, sports, and devout observances... At this as at any other cultural stage, government and war are, at least in part, carried out for the pecuniary gain of those who engage in them, but it is gain obtained by the honourable method of seizure and conversion."

Veblen also remarks that "It is true of dress even in a higher degree than of most items of consumption, that people will undergo a very considerable degree of privation in the comforts or the necessities of life in order to afford what is considered a decent amount of wasteful consumption; so that it is by no means an uncommon occurrence, in an inclement climate, for people to go ill clad in order to appear well dressed."

The sensation caused by the publication of Veblen's book, and the fact that his phrase, "conspicuous consumption", has become part of our language, indicate that his theory did not completely miss its mark. In fact, modern advertisers seem to be following Veblen's advice: Realizing that much of the output of our economy will be used for the purpose of establishing the social status of consumers, advertising agencies hire psychologists to appeal to the consumer's longing for a higher social position.

When possessions are used for the purpose of social competition, demand has no natural upper limit; it is then limited only by the size of the human ego, which, as we know, is boundless. This would be all to the good if unlimited economic growth were desirable. But today, when further growth implies future collapse, industrial society urgently needs to find new values to replace our worship of power, our restless chase after excitement, and our admiration of excessive consumption.

6.2 Gandhi as an economist

If humans are to achieve a stable society in the distant future, it will be necessary for them to become modest in their economic behavior and peaceful in their politics. For both modesty and peace, Gandhi is useful as a source of ideas.

Mohandas Karamchand Gandhi was born in 1869 in Porbandar, India. His family belonged to the Hindu caste of shopkeepers. (In Gujarati "Gandhi" means "grocer".) However, the family had risen in status, and Gandhi's father, grandfather, and uncle had all served as prime ministers of small principalities in western India.

In 1888, Gandhi sailed for England, where he spent three years studying law at the Inner Temple in London. Before he left India, his mother had made him take a solemn oath not to touch women, wine, or meat. He thus came into contact with the English vegetarians, who included Sir Edward Arnold (translator of the Bhagavad Gita), the Theosophists Madame Blavatski and Annie Besant, and the Fabians. Contact with this idealistic group of social critics and experimenters helped to cure Gandhi of his painful shyness, and it also developed his taste for social reform and experimentation.

Gandhi's exceptionally sweet and honest character won him many friends in England, and he encountered no racial prejudice at all. However, when he traveled to Pretoria in South Africa a few years later, he experienced racism in its worst form. Although he was meticulously well dressed in an English frock coat, and in possession of a first-class ticket, Gandhi was given the choice between traveling third class or being thrown off the train. (He chose the second alternative.) Later in the journey he was beaten by a coach driver because he insisted on his right to sit as a passenger rather than taking a humiliating position on the footboard of the coach.

The legal case which had brought Gandhi to South Africa was a dispute between a wealthy Indian merchant, Dada Abdullah Seth, and his relative, Seth Tyeb (who had refused to pay a debt of 40,000 pounds, in those days a huge sum). Gandhi succeeded in reconciling these two relatives, and he persuaded them to settle their differences out of court. Later he wrote about this experience:

"Both were happy with this result, and both rose in public estimation. My joy was boundless. I had learnt the true practice of law. I had learnt to find out the better side of human nature and to enter men's hearts. I realized that the true function of a lawyer was to unite parties riven asunder. The lesson was so indelibly burnt into me that a large part of my time during my twenty years of practice as a lawyer was occupied in bringing about compromises of hundreds of cases. I lost nothing thereby - not even money, certainly not my soul."

Gandhi was about to return to India after the settlement of the case, but at a farewell party given by Abdullah Seth, he learned of a bill before the legislature which would deprive Indians in South Africa of their right to vote. He decided to stay and fight against the bill.

Gandhi spent the next twenty years in South Africa, becoming the leader of a struggle for the civil rights of the Indian community. In this struggle he tried "...to find the better side of human nature and to enter men's hearts." Gandhi's stay in England had given him a glimpse of English liberalism and English faith in just laws. He felt confident that if the general public in England could be made aware of gross injustices in any part of the British Empire, reform would follow. He therefore organized non-violent protests in which the protesters sacrificed themselves so as to show as vividly as possible the injustice of an existing law. For example, when the government ruled that Hindu, Muslim and Parsi marriages had no legal standing, Gandhi and his followers voluntarily went to prison for ignoring the ruling.

Gandhi used two words to describe this form of protest: "satyagraha" (the force of truth) and "ahimsa" (non-violence). Of these he later wrote: "I have nothing new to teach the world. Truth and non-violence are as old as the hills. All that I have done is to try experiments in both on as vast a scale as I could. In so doing, I sometimes erred and learnt by my errors. Life and its problems have thus become to me so many experiments in the practice of truth and non-violence."

In his autobiography, Gandhi says: "Three moderns have left a deep impression on my life and captivated me: Raychandbhai (the Indian philosopher and poet) by his living contact; Tolstoy by his book 'The Kingdom of God is Within You'; and Ruskin by his book 'Unto This Last'."

Ruskin's book, "Unto This Last", which Gandhi read in 1904, is a criticism of modern industrial society. Ruskin believed that friendships and warm interpersonal relationships are a form of wealth that economists have failed to consider. He felt that warm human contacts are most easily achieved in small agricultural communities, and that therefore the modern tendency towards centralization and industrialization may be a step backward in terms of human happiness. While still in South Africa, Gandhi founded two religious Utopian communities based on the ideas of Tolstoy and Ruskin. Phoenix Farm (1904) and Tolstoy Farm (1910). At this time he also took an oath of chastity ("bramacharya"), partly because his wife was unwell and he wished to protect her from further pregnancies, and partly in order to devote himself more completely to the struggle for civil rights.

Because of his growing fame as the leader of the Indian civil rights movement in South Africa, Gandhi was persuaded to return to India in 1914 and to take up the cause of Indian home rule. In order to reacquaint himself with conditions in India, he traveled tirelessly, now always going third class as a matter of principle.

During the next few years, Gandhi worked to reshape the Congress Party into an organization which represented not only India's Anglicized upper middle class but also the millions of uneducated villagers who were suffering under an almost intolerable burden of poverty and disease. In order to identify himself with the poorest of India's people, Gandhi began to wear only a white loincloth made of rough homespun cotton. He traveled to the remotest villages, recruiting new members for the Congress Party, preaching non-violence and "firmness in the truth", and becoming known for his voluntary poverty and humility. The villagers who flocked to see him began to call him "Mahatma" (Great Soul).

Disturbed by the spectacle of unemployment and poverty in the villages, Gandhi urged the people of India to stop buying imported goods, especially cloth, and to make their own. He advocated the reintroduction of the spinning wheel into village life, and he often spent some hours spinning himself. The spinning wheel became a symbol of the Indian independence movement, and was later incorporated into the Indian flag.

Fig. 6.2 *Gandhi and his wife Kasturbhai in 1902.*

Fig. 6.3 *Gandhi's spinning wheel was incorporated into the flag of the Congress Party and later into the national flag of an independent India.*

The movement for boycotting British goods was called the "Swadeshi movement". The word Swadeshi derives from two Sanskrit roots: *Swa*, meaning self, and *Desh*, meaning country. Gandhi described Swadeshi as "a call to the consumer to be aware of the violence he is causing by supporting those industries that result in poverty, harm to the workers and to humans or other creatures."

Gandhi tried to reconstruct the crafts and self-reliance of village life that he felt had been destroyed by the colonial system. "I would say that if the village perishes India will perish too", he wrote, "India will be no more India. Her own mission in the world will get lost. The revival of the village is only possible when it is no more exploited. Industrialization on a mass scale will necessarily lead to passive or active exploitation of the villagers as problems of competition and marketing come in. Therefore we have to concentrate on the village being self-contained, manufacturing mainly for use. Provided this character of the village industry is maintained, there would be no objection to villagers using even the modern machines that they can make and can afford to use. Only they should not be used as a means of exploitation by others."

"You cannot build nonviolence on a factory civilization, but it can be built on self-contained villages... Rural economy as I have conceived it, eschews exploitation altogether, and exploitation is the essence of violence... We have to make a choice between India of the villages that are as ancient as herself and India of the cities which are a creation of foreign domination..."

"Machinery has its place; it has come to stay. But it must not be allowed to displace necessary human labour. An improved plow is a good thing. But if by some chances, one man could plow up, by some mechanical invention of his, the whole of the land of India, and control all the agricultural produce, and if the millions had no other occupation, they would starve, and being idle, they would become dunces, as many have already become. There is hourly danger of many being reduced to that unenviable state."

In these passages we see Gandhi not merely as a pioneer of nonviolence; we see him also as an economist. Faced with misery and unemployment produced by machines, Gandhi tells us that social goals must take precedence over blind market mechanisms. If machines are causing unemployment, we can, if we wish, and use labor-intensive methods instead. With Gandhi, the free market is not sacred - we can do as we wish, and maximize human happiness, rather than maximizing production and profits.

Gandhi also organized many demonstrations whose purpose was to show the British public that although the British raj gave India many benefits, the toll exacted was too high, not only in terms of money, but also in terms of India's self-respect and self-sufficiency. All of Gandhi's demonstrations were designed to underline this fact. For example, in 1930 Gandhi organized a civil-disobedience campaign against the salt laws. The salt laws gave the Imperial government a monopoly and prevented Indians from making their own salt by evaporating sea water. The majority of Indians were poor farmers who worked long hours in extreme heat, and salt was as much a necessity to them as bread. The tax on salt was essentially a tax on the sweat of the farmers.

Before launching his campaign, Gandhi sent a polite letter to the Viceroy, Lord Irwin, explaining his reasons for believing that the salt laws were unjust, and announcing his intention of disregarding them unless they were repealed. Then, on March 12 1930, Gandhi and many of his followers, accompanied by several press correspondents, started on a march to the sea to carry out their intention of turning themselves into criminals by making salt. Every day, Gandhi led the procession about 12 miles, stopping at villages in the evenings to hold prayer meetings. Many of the villagers joined the march, while others cast flower petals in Gandhi's path or sprinkled water on his path to settle the dust.

On April 5 the marchers arrived at the sea, where they spent the night in prayer on the beach. In the morning they began to make salt by wading into the sea, filling pans with water, and letting it evaporate in the sun. Not much salt was made in this way, but Gandhi's action had a strong symbolic power. A wave of non-violent civil disobedience demonstrations swept over India, so extensive and widespread that the Imperial government, in danger of losing control of the country, decided to arrest as many of the demonstrators as possible. By midsummer, Gandhi and a hundred thousand of his followers were in prison, but nevertheless the civil disobedience demonstrations continued.

In January, 1931, Gandhi was released from prison and invited to the Viceroy's palace to talk with Lord Irwin. They reached a compromise agreement: Gandhi was to call off the demonstrations and would attend a Round Table Conference in London to discuss Indian home rule, while Lord Irwin agreed to release the prisoners and would change the salt laws so that Indians living near to the coast could make their own salt.

The salt march was typical of Gandhi's non-violent methods. Throughout the demonstrations he tried to maintain a friendly attitude towards his opponents, avoiding escalation of the conflict. Thus at the end of the demonstrations, the atmosphere was one in which a fair compromise solution could be reached. Whenever he was in prison, Gandhi regarded his jailers as his hosts. Once, when he was imprisoned in South Africa, he used the time to make a pair of sandals, which he sent to General Smuts, the leader of the South African government. Thus Gandhi put into practice the Christian principle, "Love your enemies; do good to them that hate you."

Gandhi's importance lies in the fact that he was a major political leader who sincerely tried to put into practice the ethical principles of religion. In his autobiography Gandhi says: "I can say without the slightest hesitation, and yet with all humility, that those who say that religion has nothing to do with politics do not know what religion means."

Gandhi believed that human nature is essentially good, and that it is our task to find and encourage whatever is good in the character of others. During the period when he practiced as a lawyer, Gandhi's aim was "to unite parties riven asunder," and this was also his aim as a politician. In order for reconciliation to be possible in politics, it is necessary to avoid escalation of conflicts. Therefore Gandhi used non-violent methods, relying only on the force of truth. "It is my firm conviction," he wrote, "that nothing can be built on violence."

To the insidious argument that "the end justifies the means," Gandhi answered firmly: "They say 'means are after all means'. I would say 'means are after all everything'. As the means, so the end. Indeed the Creator has given us control (and that very limited) over means, none over end. ... The means may be likened to a seed, and the end to a tree; and there is the same inviolable connection between the means and the end as there is between the seed and the tree. Means and end are convertible terms in my philosophy of life." In other words, a dirty method produces a dirty result; killing produces more killing; hate leads to more hate. But there are positive feedback loops as well as negative ones. A kind act produces a kind response; a generous gesture is returned; hospitality results in reflected hospitality. Hindus and Buddhists call this principle "the law of karma".

Gandhi believed that the use of violent means must inevitably contaminate the end achieved. Because Gandhi's methods were based on love, understanding, forgiveness and reconciliation, the non-violent revolution which he led left very little enmity in its wake. When India finally achieved its independence from England, the two countries parted company without excessive bitterness. India retained many of the good ideas which the English had brought - for example the tradition of parliamentary democracy - and the two countries continued to have close cultural and economic ties.

Mahatma Gandhi was assassinated by a Hindu extremist on January 30, 1948. After his death, someone collected and photographed all his worldly goods. These consisted of a pair of glasses, a pair of sandals and a white homespun loincloth. Here, as in the Swadeshi movement, we see Gandhi as a pioneer of economics. He deliberately reduced his possessions to an absolute minimum in order to demonstrate that there is no connection between personal merit and material goods. Like Veblen, Mahatma Gandhi told us that we must stop using material goods as a means of social competition. We must start to judge people not by what they have, but by what they are.

6.3 Thoreau

In the distant future (and perhaps even in the not-so-distant future) industrial civilization will need to abandon its relentless pursuit of unnecessary material goods and economic growth. Modern society will need to re-establish a balanced and harmonious relationship with nature. In pre-industrial societies harmony with nature is usually a part of the cultural tradition. In our own time, the same principle has become central to the ecological counter-culture while the main-stream culture thunders blindly ahead, addicted to wealth, power and growth.

In the 19th century the American writer, Henry David Thoreau (1817-1862), pioneered the concept of a simple life, in harmony with nature. Today, his classic book, *Walden*, has become a symbol for the principles of ecology, simplicity, and respect for nature.

Thoreau was born in Concord Massachusetts, and he attended Harvard from 1833 to 1837. After graduation, he returned home, worked in his family's pencil factory, did odd jobs, and for three years taught in a progressive school founded by himself and his older brother, John. When John died of lockjaw in 1842, Henry David was so saddened that he felt unable to continue the school alone.

Thoreau refused to pay his poll tax because of his opposition to the Mexican War and to the institution of slavery. Because of his refusal to pay the tax (which was in fact a very small amount) he spent a night in prison. To Thoreau's irritation, his family paid the poll tax for him and he was released. He then wrote down his ideas on the subject in an essay entitled *The Duty of Civil Disobedience*, where he maintains that each person has a duty to follow his own individual conscience even when it conflicts with the orders of his government. "Under a government that which imprisons any unjustly", Thoreau wrote, "the true place for a just man is in prison." *Civil Disobedience* influenced Tolstoy, Gandhi and Martin Luther King, and it anticipated the Nüremberg Principles.

Thoreau became the friend and companion of the transcendentalist writer Ralph Waldo Emerson (1803-1882), who introduced him to a circle of writers and thinkers that included Ellery Channing, Margaret Fuller and Nathanial Hawthorne.

Nathanial Hawthorne described Thoreau in the following words: "Mr. Thorow [sic] is a keen and delicate observer of nature - a genuine observer, which, I suspect, is almost as rare a character as even an original poet; and Nature, in return for his love, seems to adopt him as her especial child, and shows him secrets which few others are allowed to witness. He is familiar with beast, fish, fowl, and reptile, and has strange stories to tell of adventures, and friendly passages with these lower brethren of mortality. Herb and flower, likewise, wherever they grow, whether in garden, or wild wood, are his familiar friends. He is also on intimate terms with the clouds and can tell the portents of storms. It is a characteristic trait, that he has a great regard for the memory of the Indian tribes, whose wild life would have suited him so well; and strange to say, he seldom walks over a plowed field without picking up an arrow-point, a spear-head, or other relic of the red men - as if their spirits willed him to be the inheritor of their simple wealth."

At Emerson's suggestion, Thoreau opened a journal, in which he recorded his observations concerning nature and his other thoughts. Ultimately the journal contained more than 2 million words. Thoreau drew on his journal when writing his books and essays, and in recent years, many previously unpublished parts of his journal have been printed.

From 1845 until 1847, Thoreau lived in a tiny cabin that he built with his own hands. The cabin was in a second-growth forest beside Walden Pond in Concord, on land that belonged to Emerson. Thoreau regarded his life there as an experiment in simple living. He described his life in the forest and his reasons for being there in his book *Walden*, which was published in 1854. The book is arranged according to seasons, so that the two-year sojourn appears compressed into a single year.

Fig. 6.4 *Henry David Thoreau, 1817-1862.*

"Most of the luxuries", Thoreau wrote, "and many of the so-called comforts of life, are not only not indispensable, but positive hindrances to the elevation of mankind. With respect to luxuries, the wisest have ever lived a more simple and meager life than the poor. The ancient philosophers, Chinese, Hindoo, Persian, and Greek, were a class than which none has been poorer in outward riches, none so rich in inward."

Elsewhere in *Walden*, Thoreau remarks, "It is never too late to give up your prejudices", and he also says, "Why should we be in such desperate haste to succeed, and in such desperate enterprises? If a man does not keep pace with his companions, perhaps it is because he hears a different drummer." Other favorite quotations from Thoreau include "Rather than love, than money, than fame, give me truth", "Beware of all enterprises that require new clothes", "Most men lead lives of quiet desperation" and "Men have become tools of their tools."

Towards the end of his life, when he was very ill, someone asked Thoreau whether he had made his peace with God. "We never quarreled", he answered.

Thoreau's closeness to nature can be seen from the following passage, written by his friend Frederick Willis, who visited him at Walden Pond in 1847, together with the Alcott family: "He was talking to Mr. Alcott of the wild flowers in Walden woods when, suddenly stopping, he said: 'Keep very still and I will show you my family.' Stepping quickly outside the cabin door, he gave a low and curious whistle; immediately a woodchuck came running towards him from a nearby burrow. With varying note, yet still low and strange, a pair of gray squirrels were summoned and approached him fearlessly. With still another note several birds, including two crows flew towards him, one of the crows nestling upon his shoulder. I remember that it was the crow resting close to his head that made the most vivid impression on me, knowing how fearful of man this bird is. He fed them all from his hand, taking food from his pocket, and petted them gently before our delighted gaze; and then dismissed them by different whistling, always strange and low and short, each wild thing departing instantly at hearing his special signal."

In an essay published by the *Atlantic Monthly* in 1853, Thoreau described a pine tree in Maine with the words: "It is as immortal as I am, and perchance will go to as high a heaven, there to tower above me still." However, the editor (James Russell Lowell) considered the sentence to be blasphemous, and removed it from Thoreau's essay before publication.

In one of his essays, Thoreau wrote: "If a man walk in the woods for love of them half of each day, he is in danger of being regarded as a loafer; but if he spends his whole day as a speculator, shearing off those woods and making the earth bald before her time, he is esteemed an industrious and enterprising citizen."

6.4 The counter-culture

In Chapter 2, we mentioned Say's Law ("Supply creates its own demand"). Jean-Baptiste Say's basis for this proposition was the assumption that a consumer's desire for goods is infinite. He combined this assumption with the observation that the wages paid for the production of goods will provide money enough to buy back the goods, even if the amount involved increases without limit. Comforted by Say's "law", and by the observation that people in industrial societies do indeed consume far more than they actually need, economists continue to pursue economic growth as though it were the Holy Grail. We do indeed devote much of our efforts to "making the earth bald before her time".

As things are today, the advertising industry, which is part of the mainstream culture, whips demand towards ever higher levels by exploiting our tendency to use material goods for the purpose of social competition. Meanwhile, a small but significant counter-culture has realized that unlimited economic growth will lead to ecological disaster unless we stop in time.

In the 1960's, a counter-culture developed in the United States, partly as a reaction against the Vietnam War and partly as a reaction against consumerism. It seemed to young people that they were being offered a possession-centered way of life that they did not want, and that they were being asked to participate in a war that they thought was immoral.

In 1964, a free speech movement began on the campus of the University of California in Berkeley. Students demanded that the university administration should lift a ban that it had imposed on on-campus political activities. Student movements elsewhere in the United States and in Europe echoed the Berkeley protests throughout the late 1960's and early 1970's.

Mario Savo, one of the leaders of the Berkeley free speech movement, compared the Establishment to an enormous anti-human machine: "There is a time when the operation of the machine becomes so odious, makes you so sick at heart, that you can't take part; you can't even passively take part, and you've got to put your bodies upon the gears and upon the wheels, upon the levers, upon all the apparatus, and you've got to make it stop. And you've got to indicate to the people who run it, to the people who own it, that unless you're free, the machine will be prevented from working at all."

The Greening of America, by Charles Reich, describes the youth-centered counter-culture: "Industrialism produced a new man...", Reich wrote, "one adapted to the demands of the machine. In contrast, today's emerging consciousness seeks a new knowledge of what it means to be human, in order that the machine, having been built, may now be turned to human ends; in order that man once more can become a creative force, renewing and creating his own life and thus giving life back to society."

6.5 The Brundtland Report

In 1972, the United Nations Conference on the Human Environment took place in Stockholm. In a 1983 follow-up to the Stockholm conference, the General Assembly of the UN adopted a resolution (A/38/161) establishing the World Commission on Environment and Development. It is usually known as the Brundtland Commission after the name of its Chair, Dr. Gro Harlem Brundtland, who was at the time the Prime Minister of Norway. The report of the Brundtland Commission, entitled *Our Common Future*, was submitted to the United Nations in 1987.

In the words of Dr. Brundtland, the goal of the report was "to help define shared perceptions of long-term environmental issues and the appropriate efforts needed to deal successfully with the problems of protecting and enhancing the environment, a long-term agenda for action during the coming decades..."

One of the key concepts of the Brundtland Report was "sustainable development". The Report offered the following definition: "Sustainable development is development that meets the needs of the present without compromising the ability of future generations to meet their own needs."

Fig. 6.5 *Gro Harlem Brundtland*

The Brundtland Commission's key concepts for sustainability were as follows:

(1) Today's needs should not compromise the ability of future generations to meet their needs.
(2) A direct link exists between the economy and the environment.
(3) The needs of the poor in all nations must be met.
(4) In order for the environment to be protected, the economic conditions of the world's poor must be improved.
(5) In all our actions, we must consider the impact upon future generations.

The Brundtland Commission's report examines the question of whether the earth can support a population of 10 billion people without the collapse of the ecological systems on which all life depends. The report states that the data "suggest that meeting the food requirements of an ultimate world population of around 10 billion would require some changes in food habits, as well as greatly improving the efficiency of traditional agriculture."

6.6 The Earth Summit at Rio

The Brundtland Report served as a preparation for the United Nations Conference on Environment and Development, which took place from the 3rd to the 14th of June, 1992 in Rio de Janeiro. The conference, informally called the "Earth Summit", was unprecedented in its size and significance. 172 governments participated, including 108 heads of state or government. 17,000 people attended the Earth Summit, including 2,400 representatives of NGO's. An estimated 10,000 journalists covered the conference.

The Earth Summit at Rio ought to have been a turning point in the relationship between humans and the global environment. However, despite the size and importance of the conference, and despite the hopes of most of the participants, the the Earth Summit did not result in the changes in laws and lifestyles that will be needed to establish long-term sustainability.

Two basic problems are leading to the destruction of the global environment - excessive population growth in the developing South, and excessive economic growth and overconsumption in the industrial North. Political and religious pressures prevented overpopulation from being named at Rio as one of the root causes of environmental degradation. Political pressures also prevented the necessary changes in laws and lifestyles from being made in the North.

Nevertheless, considerable progress was made at Rio. The resulting documents included Agenda 21 (an environmental agenda for the 21st century), the Rio Declaration on Environment and Development, the Statement on Forest Principles, the United Nations Framework Convention on Climate Change and the United Nations Convention on Biological Diversity. Later the Earth Charter was developed by some of the leaders who met in Rio.

Agenda 21

The first few chapters of Agenda 21 are as follows:

(1) Preamble
(2) International cooperation to accelerate sustainable development in developing countries and related domestic policies
(3) Combating poverty
(4) Changing consumption patterns
(5) Demographic dynamics and sustainability
(6) Protecting and promoting human health conditions
(7) Promoting sustainable human settlement development
(8) Integrating environment and development in decision-making
(9) Protecting the atmosphere
(10) Integrated approach to the planning and management of land resources
(11) Combating deforestation
(12) Managing fragile ecosystems; sustainable mountain development
(13) Conservation of biological diversity
(14) Environmentally sound management of biotechnology
(15) Protection of the oceans

The good intentions of the authors shine from this list! It was a major victory to have Agenda 21 adopted as the official policy of the United Nations. Close examination reveals many political compromises in the wording the conclusions, but the idealism of the document is not entirely lost.

Agenda 21, touches (very lightly!) on the root causes of environmental degradation. In Section 4.6, one finds the extremely weak statement: "Some economists are questioning traditional concepts of economic growth and underlining the importance of pursuing economic objectives that take into account of the full value of natural resource capital. More needs to be known about the role of consumption in relation to economic growth and population dynamics in order to formulate coherent international and national policies." However, in Section 5.3, a clearer statement of the basic problem appears: "The growth of world population and production, combined with unsustainable consumption patterns, places increasingly severe stress on the life-supporting systems of our planet."

Suggestions for further reading

(1) R. Tilman, *The Intellectual Legacy of Thorstein Veblen: Unresolved Issues*, Greenwood Press, (1996).
(2) R. Tilman, *Thorstein Veblen and His Critics, 1891-1963*, Princeton University Press, (1992).
(3) K. McCormick, *Veblen in Plain English*, Cambria Press, (2006).
(4) J. Dorfman, *Thorstein Veblen and His America*, Harvard University Press, (1934).
(5) J. Homer, ed., *The Gandhi Reader: A Sourcebook of his Life and Writings*, Grove Press, New York, (1956).
(6) G. Sharp, *Gandhi as a Political Strategist, with Essays on Ethics and Politics*, Extending Horizon Books, Boston, (1979).
(7) J.V. Bondurant, *Conquest of Violence: The Gandhian Philosophy of Conflict*, Princeton University Press, (1988).
(8) L. Fischer, *The Essential Gandhi: An Anthology of his Writings on His Life, Work and Ideas*, Vintage, New York, (2002).
(9) M.K. Gandhi, *Hind Swaraj and Other Writings*, edited by A.J. Parel, Cambridge Texts in Modern Politics, (2006).
(10) C. Bode, *Best of Thoreau's Journals*, Southern Illinois University Press, (1967).
(11) J. Meyerson et al., *The Cambridge Companion to Henry David Thoreau*, Cambridge University Press, (1995).
(12) W. Howarth, *The Book of Concord: Thoreau's Life as a Writer*, Viking Press, (1982).
(13) W. Harding, *Days of Henry Thoreau*, Princeton University Press, (1982).
(14) T. Roszak, *The Making of a Counter Culture*, (1970).
(15) E. Nelson, *The British Counterculture 1966-1973*, Macmillan, London, (1989).
(16) G. McKay, *Senseless Acts of Beauty: Cultures of Resistance since the Sixties*, Verso, London, (1996).
(17) K. Goffman, *Counterculture Through the Ages*, Villard Books, (2004).
(18) Brundtland Commission, *Our Common Future*, Oxford University Press, (1987).
(19) G.O. Barney, , *The Unfinished Agenda: The Citizen's Policy Guide to Environmental Issues*, Thomas Y. Crowell, New York, (1977).
(20) R.E. Benedick, *Ozone Diplomacy: New Directions in Safeguarding the Planet*, Harvard University Press, Cambridge, (1991).
(21) T. Berry, *The Dream of the Earth*, Sierra Club Books, San Francisco, (1988).

(22) L.R. Brown, *The Twenty-Ninth Day*, W.W. Norton, New York, (1978).
(23) M.E. Clark, *Ariadne's Thread: The Search for New Modes of Thinking*, St. Martin's Press, New York, (1989).
(24) W.C. Clark and others, *Managing Planet Earth*, Special Issue, *Scientific American*, September, (1989).
(25) B. Commoner, *The Closing Circle: Nature, Man and Technology*, Bantam Books, New York, (1972).
(26) Council on Environmental Quality and U.S. Department of State, *Global 2000 Report to the President: Entering the Twenty-First Century*, Technical Report, Volume 2, U.S. Government Printing Office, Washington D.C., (1980).
(27) J.C.I. Dooge et al. (editors), *Agenda of Science for Environment and Development into the 21st Century*, Cambridge University Press, (1993).
(28) E. Eckholm, *The Picture of Health: Environmental Sources of Disease*, New York, (1976).
(29) Economic Commission for Europe, *Air Pollution Across Boundaries*, United Nations, New York, (1985).
(30) P.R. Ehrlich, A.H. Ehrlich and J. Holdren, *Ecoscience: Population, Resources, Environment*, W.H. Freeman, San Francisco, (1977)
(31) P.R. Ehrlich and A.H. Ehrlich, *Extinction*, Victor Gollancz, London, (1982).
(32) P.R. Ehrlich and A.H. Ehrlich, *Healing the Planet*, Addison Wesley, Reading MA, (1991).
(33) C. Flavin, *Slowing Global Warming: A Worldwide Strategy*, Worldwatch Paper 91, Worldwatch Institute, Washington D.C., (1989).
(34) H.F. French, *Clearing the Air: A Global Agenda* , Worldwatch Paper 94, Worldwatch Institute, Washington D.C., (1990).
(35) H.F. French, *After the Earth Summit: The Future of Environmental Governance*, Worldwatch Paper 107, Worldwatch Institute, Washington D.C., (1992).
(36) G. Hagman and others, *Prevention is Better Than Cure*, Report on Human Environmental Disasters in the Third World, Swedish Red Cross, Stockholm, Stockholm, (1986).
(37) G. Hardin, "The Tragedy of the Commons", *Science*, December 13, (1968).
(38) P.W. Hemily and M.N. Ozdas (eds.) *Science and Future Choice*, Clarendon, Oxford, (1979).
(39) IUCN, UNEP, WWF, *Caring for the Earth*, Earthscan Publications, London, (1991).

(40) L. Rosen and R.Glasser (eds.), *Climate Change and Energy Policy*, Los Alamos National Laboratory, AIP, New York, (1992).
(41) J.J. MacKenzie and M.T. El-Ashry, *Ill Winds: Airborne Pollution's Toll on Trees and Crops*, World Resources Institute, Washington D.C., (1988).
(42) J.T. Mathews (editor), *Preserving the Global Environment: The Challenge of Shared Leadership*, W.W. Norton, New York, (1991).
(43) J. McCormick, *Acid Earth*, International Institute for Environment and Development, London, (1985).
(44) N. Myers, *The Sinking Ark*, Pergamon, New York, (1972).
(45) N. Myers, *Conservation of Tropical Moist Forests*, National Academy of Sciences, Washington D.C., (1980).
(46) D.W. Orr, *Ecological Literacy*, State University of New York Press, Albany, (1992).
(47) D.C. Pirages and P.R. Ehrlich, *Ark II: Social Responses to Environmental Imperatives*, W.H. Freeman, San Francisco, (1974).
(48) J. Rotblat (ed.), *Shaping Our Common Future: Dangers and Opportunities (Proceedings of the Forty-Second Pugwash Conference on Science and World Affairs)*, World Scientific, London, (1994).
(49) J.C. Ryan, *Life Support: Conserving Biological Diversity*, Worldwatch Paper 108, Worldwatch Institute, Washington D.C., (1992).
(50) S.F. Singer, *Global Effects of Environmental Pollution*, Springer Verlag, New York, (1971).
(51) B. Stokes, *Local Responses to Global Problems: A Key to Meeting Basic Human Needs*, Worldwatch Paper 17, Worldwatch Institute, Washington D.C., (1978).
(52) L. Timberlake, *Only One Earth: Living for the Future*, BBC/ Earthscan, London, (1987).
(53) UNEP, *Environmental Data Report*, Blackwell, Oxford, (published annually).
(54) UNESCO, *International Coordinating Council of Man and the Biosphere*, MAB Report Series No. 58, Paris, (1985).
(55) P.M. Vitousek, P.R. Ehrlich, A.H. Ehrlich and P.A. Matson, *Human Appropriation of the Products of Photosynthesis*, Bioscience, *34*, 368-373, (1986).
(56) B. Ward and R. Dubos, *Only One Earth*, Penguin Books Ltd., (1973).
(57) P. Weber, *Abandoned Seas: Reversing the Decline of the Oceans*, Worldwatch Paper 116, Worldwatch Institute, Washington D.C., (1993).
(58) E.O. Wilson (ed.), *Biodiversity*, National Academy Press, Washington D.C., (1988).

(59) E.O. Wilson, *The Diversity of Life*, Allen Lane, The Penguin Press, London, (1992).
(60) G. Woodwell (ed.), *The Earth in Transition: Patterns and Processes of Biotic Impoverishment*, Cambridge University Press, (1990).
(61) World Commission on Environment and Development, *Our Common Future*, Oxford University Press, (1987).
(62) World Resources Institute (WRI), *Global Biodiversity Strategy*, The World Conservation Union (IUCN), United Nations Environment Programme (UNEP), (1992).

Chapter 7

GROWTH AND NONRENEWABLE RESOURCES

7.1 Biology and economics

Classical economists pictured the world as largely empty of human activities. According to the "empty-world" picture of economics, the limiting factors in the production of food and goods are shortages of capital and labor. The land, forests, fossil fuels, minerals, oceans filled with fish, and other natural resources upon which human labor and capital operate, are assumed to be present in such large quantities that they are not limiting factors. In this picture, there is no naturally-determined upper limit to the total size of the human economy. It can continue to grow as long as new capital is accumulated, as long as new labor is provided by population growth, and as long as new technology replaces labor by automation.

Biology, on the other hand, presents us with a very different picture. Biologists remind us that if any species, including our own, makes demands on its environment which exceed the environment's carrying capacity, the result is a catastrophic collapse both of the environment and of the population which it supports. Only demands which are within the carrying capacity are sustainable. For example, there is a limit to regenerative powers of a forest. It is possible to continue to cut trees in excess of this limit, but only at the cost of a loss of forest size, and ultimately the collapse and degradation of the forest. Similarly, cattle populations may for some time exceed the carrying capacity of grasslands, but the ultimate penalty for overgrazing will be degradation or desertification of the land. Thus, in biology, the concept of the carrying capacity of an environment is extremely important; but in economic theory this concept has not yet been given the weight that it deserves.

The terminology of economics can be applied to natural resources: For example, a forest can be thought of as natural capital, and the sustainable yield from the forest as interest. Exceeding the biological carrying capacity then corresponds, in economic terms, to spending one's capital. Professor Thorkil Kristensen (1899-1989), former Secretary General of the OECD, described the concept of natural capital in the following words:

"Let us try to translate pollution and ruthless exploitation of resources to economic language: Both of these mean that we are spending our capital, i.e., we are spending the earth's riches of coal, oil and raw materials, as well as our inheritance of clean air, clean water, and places where one can be free from noise pollution. It is clear that economic growth *as we experience it today* means that we spend more and more of humankind's natural wealth. This cannot continue indefinitely..."

"Instead of spending our capital, we need to begin living on our *income*. Humankind has a large, renewable source of income: *sunlight*, which constantly streams down onto our earth. It makes plants grow, upon which humans and animals can live. It does this through its action on the green leaves of plants; it maintains the cycle of water: The heat of sunlight makes the ocean water evaporate and form clouds, and drives the clouds in over our fields, where the water falls as rain and returns to the oceans through streams and rivers. The word *cycle* is symbolic. In the future, we will need to live from *natural cycles*."

It is easy to exceed the carrying capacity of an environment without realizing it. The populations of many species of wild animals exhibit oscillations which are produced when a population increases beyond the limits of sustainability and then crashes. It seems likely that the earth's population of humans is headed for a similar overshoot of the sustainable limits of its biophysical support system, followed by a crash.

There is much evidence indicating that the total size of the human economy is very rapidly approaching the absolute limits imposed by the carrying capacity of the global environment. For example, a recent study by Vitousek et al. showed that 40 percent of the net primary product of landbased photosynthesis is appropriated, directly or indirectly, for human use[1]. Thus, we are only a single doubling time away from 80 percent appropriation, which would certainly imply a disastrous degradation of the natural environment.

Another indication of our rapid approach to the absolute limit of environmental carrying capacity can be found in the present rate of loss of biodiversity. The total number of species of living organisms on the earth is thought to be between 5 million and 30 million, of which only 1.4 million have been described. Between 50% and 90% of these species live in tropical forests, a habitat which is rapidly being destroyed because of pressures

[1]The net primary product of photosynthesis is defined as the total quantity of solar energy converted to chemical energy by plants, minus the energy used by the plants themselves.

Fig. 7.1 *Sand dunes advancing towards Nouakchott, the capital of Mauritania.*

from exploding human populations. 55% of the earth's tropical forests have already been cleared and burned; and an additional area four times the size of Switzerland is lost every year. Because of this loss of habitat, tropical species are now becoming extinct at a rate which is many thousands of times the normal background rate.

If losses continue at the present rate, 20% of all tropical species will vanish irrevocably within the next 50 years One hardly dares to think of what will happen after that. The beautiful and complex living organisms on our planet are the product of more than three billion years of evolution; but today, delicately balanced and intricately interrelated communities of living things are being destroyed on a massive scale by human greed and thoughtlessness.

Further evidence that the total size of the human economy has reached or exceeded the limits of sustainability comes from global warming, from the destruction of the ozone layer, from the rate of degradation and desertification of land, from statistics on rapidly vanishing non-renewable resources, and from recent famines.

The limiting factors in economics are no longer the supply of capital or human labor or even technology. The limiting factors are the rapidly vanishing supplies of petroleum and metal ores, the forests damaged by acid rain, the diminishing catches from overfished oceans, and the cropland degraded by erosion or salination, or lost to agriculture under a cover of asphalt. Neoclassical economists have maintained that it is generally possible to substitute man-made capital for natural resources; but a closer examination shows that there are only very few cases where this is really practical.

7.2 The Club of Rome

Fig. 7.2 *A rainforest in New Zealand. Biologists estimate that between 10,000 and 50,000 species are being driven into extinction each year as the Earth's rainforests are destroyed.*

The Club of Rome is an organization of economists and scientists that was founded in 1968 by Olivetti's President, Aurelio Peccei, and the Director General for Scientific Affairs of the Organization for Economic Development and Cooperation (OECD), Alexander King. Both men were worried by the unwillingness of governments and other agencies to think carefully about the long-term future. King described the OECD in the 1960's as "a kind of temple of growth for the industrialized countries - growth for growth's sake was what mattered". Not only King and Peccei, but also Thorkil Kristensen, the Secretary General of the OECD, were worried by this veneration of growth, with little concern for its long-term consequences. They arranged for a meeting of economists and scientists to be held at the Academia dei Lincei in Rome to discuss these issues. However, they were dissatisfied with the results of the meeting because of its failure to focus on long-term problems.

After the meeting at the Academia dei Lincei, there was an informal gathering at Peccei's home, where a few like-minded participants decided to establish a private organization to study the global predicament of mankind in a perspective extending into the distant future. They agreed almost at once that the new organization should be called "The Club of Rome". Peccei described it as "an adventure of the spirit", adding that "if the Club of Rome has any merit, it is that of being the first to rebel against the suicidal ignorance of the human condition". He also stated that "It is not impossible to foster a human revolution capable of changing our course".

One of the first acts of the Club of Rome was to commission a study of future resource prospects by a group of computer scientists at the Massachusetts Institute of Technology (MIT). The group consisted of Donella H. Meadows, Dennis L. Meadows, Jorgen Randers and W.W. Behrens II. Their findings were published in 1972 in a book entitled *The Limits to Growth: A Report for the Club of Rome's Project on the Predicament of Mankind.* From the outset, the book was controversial. Many economists reacted with anger and disbelief, and even today, these feelings surface whenever *Limits to Growth* is mentioned. However, the book became a bestseller. Over 30 million copies have been sold, and it has been translated into 30 languages.

Limits to Growth distinguishes between the "static index", which its the number of years that a non-renewable resource would last at its present rate of use, and the "exponential index", defined as the number of years the resource would last at an exponentially increasing rate of use, calculated using present growth rates. Today, instead of the exponential index, the "Hubbert Peak" model for resource use has become widely accepted, and this model for the time-dependence of the depletion of a non-renewable resource will be discussed below in connection with fossil fuels.

7.3 Global energy resources

The total ultimately recoverable resources of fossil fuels amount to roughly 1260 terawatt-years of energy (1 terawatt-year $\equiv 10^{12}$ Watt-years $\equiv$ 1 TWy is equivalent to 5 billion barrels of oil or 1 billion tons of coal). Of this total amount, 760 TWy is coal, while oil and natural gas each constitute roughly 250 TWy.[2] In 1890, the rate of global consumption of energy was 1 terawatt, but by 1990 this figure had grown to 13.2 TW, distributed as follows: oil, 4.6; coal, 3.2; natural gas, 2.4; hydropower, 0.8; nuclear, 0.7; fuelwood, 0.9; crop wastes, 0.4; and dung, 0.2. By 2005, the rate of oil, natural gas and coal consumption had risen to 6.0 TW, 3.7 TW and 3.5 TW respectively. Thus, if we continue to use oil at the 2005 rate, it will last for 42 years, while

[2]British Petroleum, "B.P. Statistical Review of World Energy", London, 1991

natural gas will last for 68 years. The reserves of coal are much larger; and used at the 2005 rate, coal would last for 217 years. However, it seems likely that as oil and natural gas become depleted, coal will be converted to liquid and gaseous fuels, and its rate of use will increase. Also, the total global energy consumption is likely to increase because of increasing population and rising standards of living in the developing countries.

The industrialized countries use much more than their fair share of global resources. For example, with only a quarter of world's population they use more than two thirds of its energy; and in the U.S.A. and Canada the average per capita energy consumption is 12 kilowatts, compared with 0.2 kilowatts in Bangladesh. If we are to avoid severe damage to the global environment, the industrialized countries must rethink some of their economic ideas, especially the assumption that growth can continue forever.

Fig. 7.3 *Motor traffic in Manila.*

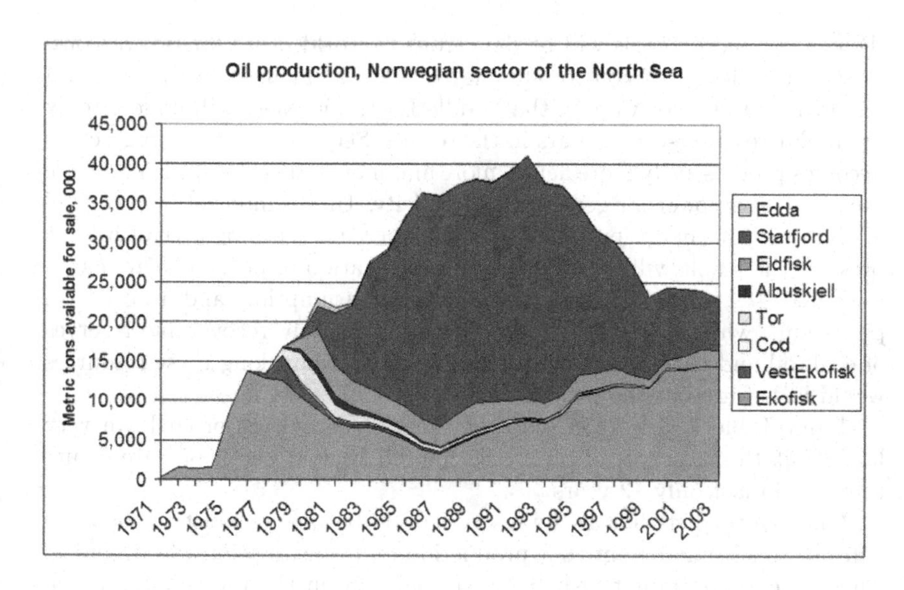

Fig. 7.4 *Local Hubbert Peaks for oil have already been reached in some regions.*

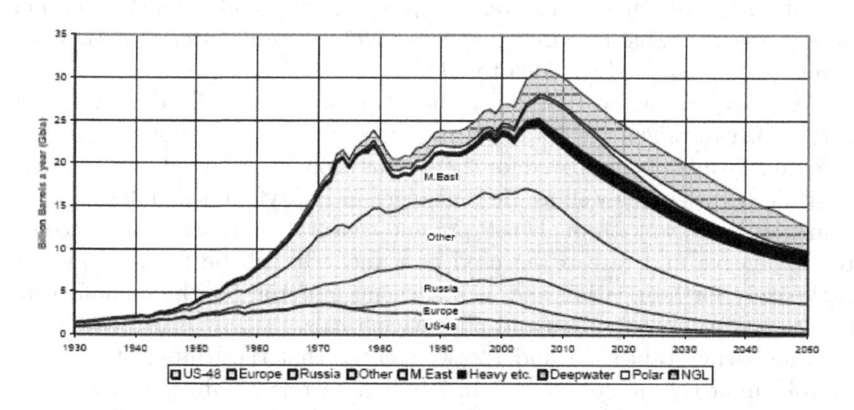

Fig. 7.5 *Production of oil (historical and projected) between 1930 and 2050. The Hubbert peak for oil is likely to occur within the next few decades. Because of rising extraction costs, the production of any non-renewable resource reaches a peak when the resource is approximately half exhausted.*

7.4 Hubbert peaks for oil and gas

Oil prices are expected to rise steadily in the future because of increasing extraction costs, vanishing reserves, and increasing demand. The effects of the coming energy crisis will be dangerous to world peace for two reasons. Firstly, the desire to control supplies of petroleum has already been an important motive for wars in the Middle East, and there will undoubtedly be similar resource-driven wars in the future. Secondly, as the energy crisis becomes progressively more acute, more and more nations will wish to make use of nuclear power for generating electricity. Unless international control of civil nuclear energy programs is made very much stronger than it is at present, the result will be wide-spread proliferation of nuclear weapons.

Tables 7.1 and 7.2 show the current consumption and reserves of petroleum, while Table 7.4 illustrates the ultimately recoverable reserves of coal, oil and natural gas, with an indication of how long these resources would last if used at the present rate.

From Table 7.4, we can see that the global reserves of coal are very large, but that reserves of oil are so limited that at the 2005 rate of use they would last only 42 years.

One can predict that as the reserves of oil become exhausted, the price will rise to such an extent that production and consumption will diminish. Thus oil experts do not visualize a special date in the future after which oil will totally disappear, but rather a date at which the production and consumption of oil will reach a maximum and afterward diminish because of scarcity of the resource and increase in price. Such a peak in the production of any nonrenewable resource is called a *Hubbert peak*, after Dr. M. King Hubbert, who applied the idea to oil reserves.

Most experts agree that the Hubbert peak for oil will occur within a decade or two. Thus the era of cheap petroleum is rapidly approaching its end, and we must be prepared for the serious economic and political impacts of rising oil prices, as well as great changes in lifestyle in the industrialized countries. Halfway through the present century, petroleum will become too expensive and rare to be used as a fuel. It will be reserved almost exclusively for lubrication and as a starting material for the manufacture of plastics, paint, fertilizers and pharmaceuticals.

Comparing Tables 7.1 and 7.2, we can see that the United States uses petroleum at the rate of more than 7 billion barrels (7 Gb) per year, while that country's estimated reserves and undiscovered resources are respectively 50.7 Gb and 49.0 Gb. Thus if the United States were to rely only on its own resources for petroleum, then, at the 2001 rate of use, these would be exhausted within 14 years. In fact, the United States already imports more than half of its oil. According to the "National Energy Policy" report (sometimes called the "Cheney Report" after its chief author) US domestic oil production will decline from 3.1 Gb/y in 2002 to 2.6 Gb/y in 2020, while US consumption will rise from 7.2 Gb/y to 9.3 Gb/y. Thus the United States today imports 57% of its oil, but the report predicts that

Table 7.1: *Oil production, reserves and resources in 1995 measured in billions of barrels (Gb). These data were originally published by Oil and Gas Journal and by US Geological Survey. 1 terawatt-year= 5.5 Gb. Only conventional petroleum is shown, i.e. superheavy forms are not included. Extraction of superheavy petroleum is very expensive.*

Country	Cumulative Production	Reserves	Undiscovered Resources	Reserves and Resources
Saudi Arabia	71.5	261.2	41.0	302.2
Iraq	22.8	112.5	45.0	157.5
Russia	92.6	100.0	68.0	168.0
Iran	42.9	93.0	22.0	115.0
UA Emirates	15.1	98.2	7.0	105.2
Kuwait	27.6	97.5	3.0	100.5
Venezuela	47.3	83.3	17.0	100.3
United States	165.8	50.7	49.0	99.7
Mexico	20.5	50.4	37.0	87.4
China	18.8	24.0	48.0	72.0
Kazakhstan	3.2	17.3	26.0	43.3
Canada	16.1	5.1	33.0	38.1
Libya	19.0	22.8	8.0	30.8
Nigeria	15.5	17.9	9.0	26.9
Norway	6.3	11.3	13.0	24.3
Indonesia	15.2	5.8	10.0	15.8
United Kingdom	12.3	4.6	11.0	15.6
Algeria	9.1	9.2	2.0	11.2
Totals	**621.6**	**1052.3**	**449.0**	**1513.8**

by 2020 this will rise to 72%. The predicted increment in US imports of oil between 2002 and 2020 is greater than the present oil consumption of China.

It is clear from these figures that if the United States wishes to maintain its enormous rate of petroleum use, it will have to rely on imported oil, much of it coming from regions of the world that are politically unstable, or else unfriendly to America, or both.

Table 7.2: *Main users of petroleum. (US Energy Information Agency, 2001.) The per-capita use of oil in China and India is very small, but it is expected to rise sharply during the next few decades.*

Country	Yearly use in billions of barrels	Population (millions)	Per-capita yearly use in barrels
United States	7.17	276	26.0
China	1.82	1262	1.4
Germany	1.03	83	12.4
Japan	0.90	127	7.1
India	0.78	1014	0.8
France	0.74	59	12.5
Mexico	0.71	100	7.1
Canada	0.70	31	22.6
Italy	0.68	58	11.7
United Kingdom	0.63	60	10.5

As the per-capita oil consumption of India and China increases, global production will fail to meet demand. For example, if the consumption in these two countries were to increase to 12 barrels per person per year (half the North American level), it would amount to 27 billion barrels per year - roughly the same amount of oil that the whole world uses today. Even a smaller increase in petroleum use by China and India may soon produce an energy crisis. One can anticipate that many voices will then be raised favoring widespread use of nuclear energy. The great dangers associated with such a development will be discussed in Appendix B.

7.5 Oilsands, tarsands and heavy oil

When the Hubbert peak for conventional oil has been passed, the price of oil will steadily increase, and this will make the extraction of oil from unconventional sources more economically feasible. For example, very large deposits of oilsands and tarsands exist in northern Alberta, Canada, a few miles north of Fort McMurray. These deposits, known as the Athabasca oil sands, consist of sand layers near to the surface. Each grain of sand in

Table 7.3: *World marketed energy demand (U.S. Energy Information Administration) measured in quadrillion BTU per year and in terawatts. The figures do not include fuelwood, dung, and other non-marketed sources of energy. World population in billions is also shown. Notice that only a small change in kilowatts per capita is projected - from 2.23 kW in 2003 to 2.93 kW in 2030. Thus the projections may underestimate future energy demand because of rapidly increasing demand from developing countries.*

year	world demand in quads/y	world demand in terawatts	world pop. in bil.	per capita kW	historical or projected
1980	283 q.	9.48 TW	4.45 b.	2.13 kW	historical
1985	309 q.	10.3 TW	4.84 b.	2.11 kW	historical
1990	347 q.	11.6 TW	5.27 b.	2.20 kW	historical
1995	366 q.	12.3 TW	5.68 b.	2.16 kW	historical
2003	421 q.	14.1 TW	6.30 b.	2.23 kW	historical
2010	510 q.	17.1 TW	6.84 b.	2.50 kW	projected
2015	563 q.	18.9 TW	7.23 b.	2.58 kW	projected
2020	613 q.	20.5 TW	7.61 b.	2.70 kW	projected
2025	665 q.	22.3 TW	7.96 b.	2.82 kW	projected
2030	722 q.	24.2 TW	8.30 b.	2.93 kW	projected

Table 7.4: *Ultimately recoverable coal, oil and natural gas reserves in 2005. 1 TWy = 10^{12} Watt-year = 5.5 billion barrels of oil = 0.76 billion tons of coal. (From BP Statistical Review of World Energy, London).*

	Global reserves	2005 global rate of consumption	Years left at 2005 rate of use
Coal	760 TWy	3.5 TW	217 years
Oil	250 TWy	6.0 TW	42 years
Natural gas	250 TWy	3.7 TW	68 years
Total	1260 TWy	13.2 TW	(95 years)

these deposits is surrounded by a thin film of water, outside of which there is a coating of oil. During the extraction process, the sand is transported to tanks where oil is stripped away from the grains by a hot water flotation process. The oil recovered in this way is too viscous to be pumped, but it can be upgraded to a pumpable fluid by the addition of naphtha. Besides the Athabasca deposit, whose area is twice the size of Lake Ontario, Alberta also has three other smaller oilsand deposits.

The energy inputs for extraction of oil from oilsands are high. It has been estimated that three barrels of oil in the sands can produce only one net barrel of output oil, because the other two barrels are needed to supply energy for the extraction process.

The world's largest deposit of superheavy oil is the "cinturon de la brea" (belt of tar) in Venezuela. This semi-solid material can be made more fluid by the addition of hydrogen. Alternatively it can be emulsified, and the emulsion can be burned in power plants.

The extremely large deposits of unconventional oil in Canada and Venezuela will to some extent cushion the economic shocks produced by

Table 7.5: *US ultimately recoverable reserves of oil, and domestic consumption (in 2001), are shown for comparison. If the US used only its domestic oil, its reserves would soon be exhausted. However, the United States imports much of its petroleum from the Middle East.*

	US reserves	2001 US rate of consumption	Years left at 2001 rate of use
Oil	**20 TWy**	**1.4 TW**	**14 years**

scarcity of conventional oil. Nevertheless, because of the high extraction costs of unconventional oil, we must still anticipate that the price of oil will rise steadily after the Hubbert peak has been reached.

7.6 Coal

The remaining reserves of coal in the world amount to about 1 exagram, i.e. 10^{18} grams or 10^{12} metric tons. The average energy density of coal is 760 Watt-years/ton, and therefore the world's coal reserves correspond to 760 TWy. If coal continues to be consumed at the present rate of 3.5 TW, the global reserves will last a little more than two centuries. However, it seems likely that as petroleum becomes prohibitively expensive, coal will be converted into liquid fuels, so that the rate of use of coal will increase. Therefore it is more realistic to lump all fossil fuels together and to divide the total supply (1260 TWy) by the the total rate of use (13.2 TW). The result is a prediction that the era of inexpensive fossil fuels will end in less than a century, as is shown in Table 7.4.

67% of the world recoverable reserves of coal are located in four countries:

(1) United States, 27%
(2) Russia, 17%
(3) China, 13%
(4) India, 10%

The present rate of use of coal by China and India is 1.5 billion metric tons per year which is equal to 1.1 TW. However, the rate of coal use by China and India is expected to double by 2030.

7.7 Climate change

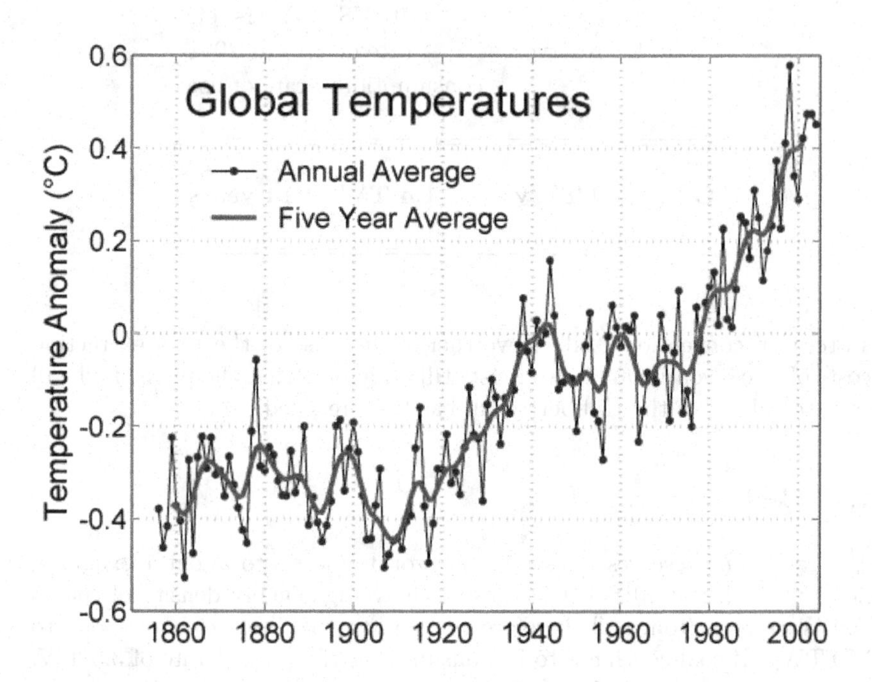

Fig. 7.6 *Global temperatures during the last two centuries.*

The burning of fossil fuels and the burning of tropical rain forests have released so much carbon dioxide that the atmospheric concentration of CO_2 has increased from a preindustrial value of 260 parts per million to 380 parts per million in 2000. At present the amount of carbon in the atmosphere is increasing by about 6 gigatons per year because of human activities; and projections estimate that the CO_2 concentration will reach about 600 ppm by 2050 (more than double the preindustrial concentration). In addition to CO_2, much methane, CH_4, is also released into the atmosphere by human activities. Anthropogenic methane comes from the production and transportation of coal, natural gas and oil, decomposition of organic wastes in municipal landfills, cultivation of rice paddies, and the raising of livestock.

The greenhouse gasses (which include water vapor, carbon dioxide, methane, ozone, nitrous oxide, sulfur hexafluoride, hydroflurocarbons, per-

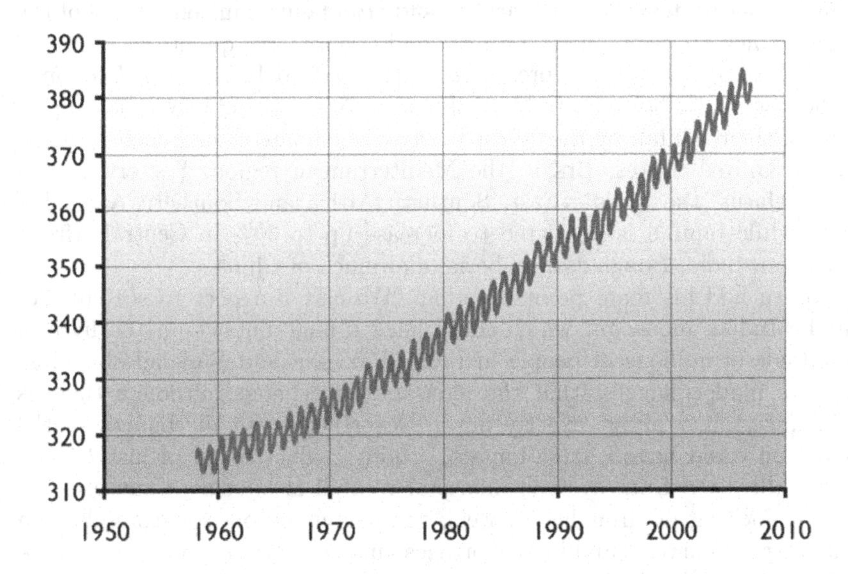

Fig. 7.7 *Atmospheric CO_2 concentrations measured at Mauna Loa, Hawaii.*

flurocarbons and many other gasses) absorb a part of the infrared radiation from the earth's surface, which otherwise would have been sent directly into outer space. Part of this energy is reradiated into space, but a part is sent downward to the earth, where it is absorbed. The result is that the earth's surface is much warmer than it otherwise would be. The mechanism is much the same as that of a greenhouse, where the glass absorbs and reradiates infrared radiation. A moderate greenhouse effect on earth is helpful to life, but climatologists believe that anthropogenic CO_2 and CH_4 emissions may produce a dangerous amount of global warming during the next few centuries.

Models put forward by the Intergovernmental Panel on Climate Change (IPCC) suggest that if no steps are taken to reduce carbon emissions, a temperature increase of 1.4-5.6 degrees C will occur by 2100[3]. Global warming may have some desirable effects, such as increased possibilities for agriculture in Canada, Sweden and Siberia. However, most of the expected effects of global warming will be damaging. These unwanted effects include ocean level rises, extreme weather conditions (such as heat waves, hurricanes and tropical cyclones), changes in the patterns of ocean currents, melting of polar ice and glaciers, abnormal spread of diseases, extinctions of plant and

[3]relative to 1990 temperatures.

animal species, together with aridity and crop failures in some areas of the world which are now able to produce and export large quantities of grain.

According to a report presented to the Oxford Institute of Economic Policy by Sir Nicholas Stern on 31 January, 2006, areas likely to lose up to 30% of their rainfall by the 2050's because of climate change include much of the United States, Brazil, the Mediterranean region, Eastern Russia and Belarus, the Middle East, Southern Africa and Southern Australia. Meanwhile rainfall is predicted to increase up to 30% in Central Africa, Pakistan, India, Bangladesh, Siberia, and much of China.

Stern and his team point out that "We can... expect to see changes in the Indian monsoon, which could have a huge impact on the lives of hundreds of millions of people in India, Pakistan and Bangladesh. Most climate models suggest that the monsoon will change, although there is still uncertainty about exactly how. Nevertheless, small changes in the monsoon could have a huge impact. Today, a fluctuation of just 10% in either direction from average monsoon rainfall is known to cause either severe flooding or drought. A weak summer monsoon, for example, can lead to poor harvests and food shortages among the rural population - two-thirds of India's almost 1.1 billion people. Heavier-than-usual monsoon downpours can also have devastating consequences..."

In some regions, melting of glaciers can be serious from the standpoint of dry-season water supplies. For example, melts from glaciers in the Hindu Kush and the Himalayas now supply much of Asia, including China and India, with a dry-season water supply. Complete melting of these glacial systems would cause an exaggerated runoff for a few decades, after which there would be a drying out of some of the most densely populated regions of the world.

The Discussion Paper presented by Stern on January 31, 2006, also notes that "Some of the potential risks could be irreversible and accelerate the process of global warming. Melting of permafrost in the Arctic could lead to the release of huge quantities of methane. Dieback of the Amazon forest could mean that the region starts to emit rather than absorb greenhouse gases. These feedbacks could lead to warming that is at least twice as fast as current high-emissions projections, leading to temperatures higher than seen in the past 50 million years. There are still uncertainties about how much warming would be needed to trigger these abrupt changes. Nevertheless, the consequences would be catastrophic if they do occur."

The much larger (700 page) Stern Report was made public on October 30, 2006. It explores not only the scientific basis for predictions of global warming but also the possible economic consequences. Unless we act promptly to prevent it, the Stern Report states, global warming could

Fig. 7.8 *Melting of Greenland's inland ice may begin in earnest in the 22nd century. When completed, this melting could cause a sea level rise of up to 7 meters. (Photo by Steve Jurvetson.)*

render swaths of the planet uninhabitable, and do economic damage equal to that inflicted by the two world wars.

In thinking about global warming, it is important to remember that it is a very slow and long-term phenomenon. Stephen H. Schneider and Janica Lane of Stanford University, in an article entitled *An Overview of 'Dangerous' Climate Change* include a figure that emphasizes the long-term nature of global warming. The figure (2.10) presupposes that CO_2 emis-

Fig. 7.9 *Two images of Mount Kilmanjaro, made on February 17, 1993 (top) and February 21, 2000 (bottom).*

sions will peak within 50 years and will thereafter be reduced. According to the figure, it will still take more than a century for the level of CO_2 in the atmosphere to stabilize. The establishment of temperature equilibrium will require several centuries. Sea level rises due to thermal expansion of ocean water will not be complete before the end of the millennium, while sea level rises due to melting of the polar icecaps might not be complete for several millennia!

It is worrying to think that total melting of the Greenland ice cap, which some authors think might begin in earnest during the 22nd century, would result in a sea level rise of up to 7 meters. Of course, society would have some time to adjust to this event. But a glance at maps and elevations makes one realize the extent of such a catastrophe and the importance of preventing it.

In conclusion, we can say that both because of limited reserves and because of the greenhouse effect, we will be forced to replace fossil fuels by renewable energy sources, and this should be done as rapidly as possible.

7.8 Metals

W. David Menzie (Chief of the Minerals Information Team of the U.S. Geological Survey) testified to a committee of the U.S. House of Representatives in 2006 that global reserves of copper are approximately 470 million tons. He also stated that world consumption of copper in 2000 was 14.9 million tons per year, but that it is increasing at 3.1% per year and is expected to reach 27 Mt/y by 2020. Menzie predicted that most of this increase will be in the developing countries. For example, China's use of copper is expected to increase from 2 Mt/y in 2000 to 5.6 Mt/y in 2020, while for India, the increase will be from 0.4 Mt/y to 1.6 Mt/y.

At the 2000 rate of use, global copper reserves will be exhausted in 31 years, while if used at a higher rate, the reserves will last for a shorter time. It is predicted that a Hubbert peak will occur for copper, analogous to the Hubbert peaks for petroleum and natural gas. Thus, copper will not disappear entirely, but there will be a date when the production of copper will reach a maximum and afterward decline because of rising prices.

The reserve index of a metal is defined as the size of its reserves divided by the current annual rate of production. Today, many metals have reserve indices between 10 years and 100 years. These include indium, tantalum, gold, bismuth, silver, cadmium, cobalt, arsenic, tungsten, molybdenum, tin, nickel, lead, zinc, and copper, while magnesium and iron have reserve indices of approximately 100 years[4].

[4]Craig, J.R., Vaugn, D.J. and Skinner, B.J., *Resources of the Earth: Origin, Use and Environmental Impact, Third Edition*, page 64.

Future exploration may increase the size of known reserves of metals; and future advances in technology may also make it possible to use lower grade ores. However, we must remember that the extraction of metals from their ores requires much energy. In the long-term future, energy will probably not be available for the production of (for example) iron, steel, and aluminum on the scale that we know today. Thus, recycling will assume great importance.

Reserves of uranium will be discussed in Appendix B.

7.9 Groundwater

It may seem surprising that fresh water can be regarded as a non-renewable resource. However, groundwater in deep aquifers is often renewed very slowly. Sometimes renewal requires several thousand years. When the rate of withdrawal of groundwater exceeds the rate of renewal, the carrying capacity of the resource has been exceeded, and withdrawal of water becomes analogous to mining a mineral. However, it is more serious than ordinary mining because water is such a necessary support for life.

In many regions of the world today, groundwater is being withdrawn faster than it can be replenished, and important aquifers are being depleted. In China, for example, groundwater levels are falling at an alarming rate. Considerations of water supply in relation to population form the background for China's stringent population policy.

At a recent lecture, Lester Brown of the Worldwatch Institute was asked by a member of the audience to name the resource for which shortages would most quickly become acute. Most of the audience expected him to name oil, but instead he replied "water". Lester Brown then cited China's falling water table. He predicted that within decades, China would be unable to feed itself. He said that this would not cause hunger in China itself: Because of the strength of China's economy, the country would be able to purchase grain on the world market. However Chinese purchases of grain would raise the price, and put world grain out of reach of poor countries in Africa. Thus water shortages in China will produce famine in parts of Africa, Brown predicted.

Under many desert areas of the world are deeply buried water tables formed during glacial periods when the climate of these regions was wetter. These regions include the Middle East and large parts of Africa. Water can be withdrawn from such ancient reservoirs by deep wells and pumping, but only for a limited amount of time.

In oil-rich Saudi Arabia, petroenergy is used to drill wells for ancient water and to bring it to the surface. Much of this water is used to irrigate wheat fields, and this is done to such an extent that Saudi Arabia exports wheat. The country is, in effect, exporting its ancient heritage of water, a policy that it may, in time, regret. A similarly short-sighted project is Muammar Gaddafi's enormous pipeline, which will bring water from ancient sub-desert reservoirs to coastal cities of Libya.

In the United States, the great Ogallala aquifer is being overdrawn. This aquifer is an enormous stratum of water-saturated sand and gravel underlying parts of northern Texas, Oklahoma, New Mexico, Kansas, Colorado, Nebraska, Wyoming and South Dakota. The average thickness of the aquifer is about 70 meters. The rate of water withdrawal from the aquifer exceeds the rate of recharge by a factor of eight.

Thus we can see that in many regions, the earth's present population is living on its inheritance of water, rather than its income. This fact, coupled with rapidly increasing populations and climate change, may contribute to a food crisis partway through the 21st century. We will discuss this danger in more detail in Chapter 10.

7.10 Topsoil

Besides depending on an adequate supply of water, food production also depends on the condition of the thin layer of topsoil that covers the world's croplands. This topsoil is being degraded and eroded at an alarming rate: According to the World Resources Institute and the United Nations Environment Programme, "It is estimated that since World War II, 1.2 billion hectares... has suffered at least moderate degradation as a result of human activity. This is a vast area, roughly the size of China and India combined." This area is 27% of the total area currently devoted to agriculture [5]. The report goes on to say that the degradation is greatest in Africa.

The risk of topsoil erosion is greatest when marginal land is brought into cultivation, since marginal land is usually on steep hillsides which are vulnerable to water erosion when wild vegetation is removed.

David Pimentel and his associates at Cornell University pointed out in 1995 that "Because of erosion-associated loss of productivity and population growth, the per capita food supply has been reduced over the past 10 years and continues to fall. The Food and Agricultural Organization reports that the per capita production of grains which make up 80% of the world's food supply, has been declining since 1984."

Pimentel et al. add that "Not only is the availability of cropland per

[5]The total area devoted to agriculture throughout the world is 1.5 billion hectares of cropland and 3.0 billion hectares of pasturage.

capita decreasing as the world population grows, but arable land is being lost due to excessive pressure on the environment. For instance, during the past 40 years nearly one-third of the world's cropland (1.5 billion hectares) has been abandoned because of soil erosion and degradation. Most of the replacement has come from marginal land made available by removing forests. Agriculture accounts for 80% of the annual deforestation."

Topsoil can also be degraded by the accumulation of salt when irrigation water evaporates. The worldwide area of irrigated land has increased from 8 million hectares in 1800 to more than 100 million hectares today. This land is especially important to the world food supply because it is carefully tended and yields are large in proportion to the area. To protect this land from salination, it should be irrigated in such a way that evaporation is minimized.

Finally cropland with valuable topsoil is being be lost to urban growth and highway development, a problem that is made more severe by growing populations and by economic growth.

Suggestions for further reading

(1) A. Gore, *An Inconvenient Truth: The Planetary Emergency of Global Warming and What We Can Do About It*, Rodale Books, New York, (2006).
(2) A. Gore, *Earth in the Balance: Forging a New Common Purpose*, Earthscan, (1992).
(3) A.H. Ehrlich and P.R. Ehrlich, *Earth*, Thames and Methuen, (1987).
(4) P.R. Ehrlich and A.H. Ehrlich, *The Population Explosion*, Simon and Schuster, (1990).
(5) P.R. Ehrlich and A.H. Ehrlich, *Healing the Planet: Strategies for Resolving the Environmental Crisis*, Addison-Wesley, (1991).
(6) P.R. Ehrlich and A.H. Ehrlich, *Betrayal of Science and Reason: How Anti-Environmental Rhetoric Threatens our Future*, Island Press, (1998).
(7) P.R. Ehrlich and A.H. Ehrlich, *One With Nineveh: Politics, Consumption and the Human Future*, Island Press, (2004).
(8) D.H. Meadows, D.L. Meadows, J. Randers, and W.W. Behrens III, *The Limits to Growth: A Report for the Club of Rome's Project on the Predicament of Mankind*, Universe Books, New York, (1972).
(9) D.H. Meadows et al., *Beyond the Limits. Confronting Global Collapse and Envisioning a Sustainable Future*, Chelsea Green Publishing, Post Mills, Vermont, (1992).
(10) D.H. Meadows, J. Randers and D.L. Meadows, *Limits to Growth: the 30-Year Update*, Chelsea Green Publishing, White River Jct., VT 05001, (2004).

(11) A. Peccei and D. Ikeda, *Before it is Too Late*, Kodansha International, Tokyo, (1984).
(12) V.K. Smith, ed., *Scarcity and Growth Reconsidered*, Johns Hopkins University Press, Baltimore, (1979).
(13) British Petroleum, *BP Statistical Review of World Energy*, (published yearly).
(14) R. Costannza, ed., *Ecological Economics: The Science and Management of Sustainability*, Colombia University Press, New York, (1991).
(15) J. Darmstadter, *A Global Energy Perspective*, Sustainable Development Issue Backgrounder, Resources for the Future, (2002).
(16) D.C. Hall and J.V. Hall, *Concepts and Measures of Natural Resource Scarcity*, *Journal of Environmental Economics and Management*, **11**, 363-379, (1984).
(17) M.K. Hubbert, *Energy Resources*, in *Resources and Man: A Study and Recommendations*, Committee on Resources and Man, National Academy of Sciences, National Research Council, W.H. Freeman, San Francisco, (1969).
(18) IPCC, Intergovernmental Panel on Climate Change, *Climate Change 2001: The Scientific Basis*, (1001).
(19) J.A. Krautkraemer, *Nonrenewable Resource Scarcity*, *Journal of Economic Literature*, bf 36, 2065-2107, (1998).
(20) N. Stern et al., *The Stern Review*, www.sternreview.org.uk, (2006).
(21) T.M. Swanson, ed., *The Economics and Ecology of Biodiversity Decline: The Forces Driving Global Change*, Cambridge University Press, (1995).
(22) P.M. Vitousek, H.A. Mooney, J. Lubchenco and J.M. Melillo, *Human Domination of Earth's Ecosystems*, *Science*, **277**, 494-499, (1997).
(23) World Resources Institute, *World Resources 200-2001: People and Ecosystems: The Fraying Web of Life*, WRI, Washington D.C., (2000).
(24) A. Sampson, *The Seven Sisters: The Great Oil Companies of the World and How They Were Made*, Hodder and Staughton, London, (1988).
(25) D. Yergin, *The Prize*, Simon and Schuster, New York, (1991).
(26) M.B. Stoff, *Oil, War and American Security: The Search for a National Policy on Oil, 1941-1947*, Yale University Press, New Haven, (1980).
(27) J. Stork, *Middle East Oil and the Energy Crisis*, Monthly Review, New York, (1976).
(28) F. Benn, *Oil Diplomacy in the Twentieth Century*, St. Martin's Press, New York, (1986).
(29) K. Roosevelt, *Countercoup: The Struggle for the Control of Iran*, McGraw-Hill, New York, (1979).

(30) E. Abrahamian, *Iran Between Two Revolutions*, Princeton University Press, Princeton, (1982).
(31) J.M. Blair, *The Control of Oil*, Random House, New York, (1976).
(32) M.T. Klare, *Resource Wars: The New Landscape of Global Conflict*, Owl Books reprint edition, New York, (2002).
(33) H. Mejcher, *Imperial Quest for Oil: Iraq, 1910-1928*, Ithaca Books, London, (1976).
(34) P. Sluglett, *Britain in Iraq, 1914-1932*, Ithaca Press, London, (1976).
(35) D.E. Omissi, *British Air Power and Colonial Control in Iraq, 1920-1925*, Manchester University Press, Manchester, (1990).
(36) V.G. Kiernan, *Colonial Empires and Armies, 1815-1960*, Sutton, Stroud, (1998).
(37) R. Solh, *Britain's 2 Wars With Iraq*, Ithaca Press, Reading, (1996).
(38) D. Morgan and D.B. Ottaway, *In Iraqi War Scenario, Oil is Key Issue as U.S. Drillers Eye Huge petroleum Pool*, Washington Post, September 15, (2002).
(39) C.J. Cleveland, *Physical and Economic Aspects of Natural Resource Scarcity: The Cost of Oil Supply in the Lower 48 United States 1936-1987*, Resources and Energy **13**, 163-188, (1991).
(40) C.J. Cleveland, *Yield Per Effort for Additions to Crude Oil Reserves in the Lower 48 States, 1946-1989*, American Association of Petroleum Geologists Bulletin, **76**, 948-958, (1992).
(41) M.K. Hubbert, *Technique of Prediction as Applied to the Production of Oil and Gas*, in *NBS Special Publication 631*, US Department of Commerce, National Bureau of Standards, (1982).
(42) L.F. Ivanhoe, *Oil Discovery Indices and Projected Discoveries*, Oil and Gas Journal, **11**, 19, (1984).
(43) L.F. Ivanhoe, *Future Crude Oil Supplies and Prices*, Oil and Gas Journal, July 25, 111-112, (1988).
(44) L.F. Ivanhoe, *Updated Hubbert Curves Analyze World Oil Supply*, World Oil, November, 91-94, (1996).
(45) L.F. Ivanhoe, *Get Ready for Another Oil Shock!*, The Futurist, January-February, 20-23, (1997).
(46) Energy Information Administration, *International Energy Outlook, 2001*, US Department of Energy, (2001).
(47) Energy Information Administration, *Caspian Sea Region*, US Department of Energy, (2001).
(48) National Energy Policy Development Group, *National Energy Policy*, The White House, (2004). (http://www.whitehouse.gov/energy/)
(49) M. Klare, *Bush-Cheney Energy Strategy: Procuring the Rest of the World's Oil*, Foreign Policy in Focus, (Interhemispheric Resource Center/Institute for Policy Studies/SEEN), Washington DC and Silver City NM, January, (2004).

(50) IEA, *CO2 from Fuel Combustion Fact-Sheet*, International Energy Agency, (2005).
(51) H. Youguo, *China's Coal Demand Outlook for 2020 and Analysis of Coal Supply Capacity*, International Energy Agency, (2003).
(52) R.H. Williams, *Advanced Energy Supply Technologies*, in *World Energy Assessment: Energy and the Challenge of Sustainability*, UNDP, (2000).
(53) H. Lehmann, *Energy Rich Japan*, Institute for Sustainable Solutions and Innovations, Achen, (2003).
(54) D. King, *Climate Change Science: Adapt, Mitigate or Ignore*, Science, **303** (5655), pp. 176-177, (2004).
(55) S. Connor, *Global Warming Past Point of No Return*, The Independent, (116 September, 2005).
(56) D. Rind, *Drying Out the Tropics*, New Scientist (6 May, 1995).
(57) J. Patz et al., *Impact of Regional Climate Change on Human Health*, Nature, (17 November, 2005).
(58) M. McCarthy, *China Crisis: Threat to the Global Environment*, The Independent, (19 October, 2005).
(59) L.R. Brown, *The Twenty-Ninth Day*, W.W. Norton, New York, (1978).
(60) W.V. Chandler, *Materials Recycling: The Virtue of Necessity*, Worldwatch Paper 56, Worldwatch Institute, Washington D.C, (1983).
(61) W.C. Clark and others, *Managing Planet Earth*, Special Issue, *Scientific American*, September, (1989).
(62) B. Commoner, *The Closing Circle: Nature, Man and Technology*, Bantam Books, New York, (1972).
(63) C. Flavin, *Slowing Global Warming: A Worldwide Strategy*, Worldwatch Paper 91, Worldwatch Institute, Washington D.C., (1989).
(64) J.R. Frisch, *Energy 2000-2020: World Prospects and Regional Stresses*, World Energy Conference, Graham and Trotman, (1983).
(65) J. Gever, R. Kaufmann, D. Skole and C. Vorosmarty, *Beyond Oil: The Threat to Food and Fuel in the Coming Decades*, Ballinger, Cambridge MA, (1986).
(66) J. Holdren and P. Herrera, *Energy*, Sierra Club Books, New York, (1971).
(67) N. Myers, *The Sinking Ark*, Pergamon, New York, (1972).
(68) National Academy of Sciences, *Energy and Climate*, NAS, Washington D.C., (1977).
(69) W. Ophuls, *Ecology and the Politics of Scarcity*, W.H. Freeman, San Francisco, (1977).
(70) A. Peccei, *The Human Quality*, Pergamon Press, Oxford, (1977).

(71) A. Peccei, *One Hundred Pages for the Future*, Pergamon Press, New York, (1977).
(72) E. Pestel, *Beyond the Limits to Growth*, Universe Books, New York, (1989).
(73) C. Pollock, *Mining Urban Wastes: The Potential for Recycling*, Worldwatch Paper 76, Worldwatch Institute, Washington D.C., (1987).
(74) S.H. Schneider, *The Genesis Strategy: Climate and Global Survival*, Plenum Press, (1976).
(75) P.B. Smith, J.D. Schilling and A.P. Haines, *Introduction and Summary*, in *Draft Report of the Pugwash Study Group: The World at the Crossroads*, Berlin, (1992).
(76) World Resources Institute, *World Resources*, Oxford University Press, New York, (published annually).
(77) J.E. Young, John E., *Mining the Earth* , Worldwatch Paper 109, Worldwatch Institute, Washington D.C., (1992).
(78) J.R. Craig, D.J. Vaughan and B.J. Skinner, *Resources of the Earth: Origin, Use and Environmental Impact, Third Edition*, Prentice Hall, (2001).
(79) W. Youngquist, *Geodestinies: The Inevitable Control of Earth Resources Over Nations and Individuals*, National Book Company, Portland Oregon, (1997).
(80) M. Tanzer, *The Race for Resources. Continuing Struggles Over Minerals and Fuels*, Monthly Review Press, New York, (1980).
(81) C.B. Reed, *Fuels, Minerals and Human Survival*, Ann Arbor Science Publishers Inc., Ann Arbor Michigan, (1975).
(82) A.A. Bartlett, *Forgotten Fundamentals of the Energy Crisis*, American Journal of Physics, **46**, 876-888, (1978).
(83) N. Gall, *We are Living Off Our Capital*, Forbes, September, (1986).

Chapter 8

RENEWABLE ENERGY

8.1 Beyond the fossil fuel era

From Table 7.4, it can be seen that fossil fuel era will be over in about a century. As oil becomes scarce, it is likely that coal will be converted to liquid fuels, as was done in Germany during World War II, and in South Africa during the oil embargo. In this process, coal is gasified to form syngas, which is a mixture of CO and H_2. These two gasses are then converted to light hydrocarbons by means of Fischer-Tropsch catalysts. Both gasoline and diesel fuel can be made in this way.

If coal is converted to liquid fuels on a large scale, the rate of use of coal will increase. Thus the projected date for the exhaustion of coal reserves based on the present consumption of coal is unrealistic. It is more accurate to lump all fossil fuels together and to predict a future date for their exhaustion based on the lumped consumption of coal, natural gas and oil. In Table 7.4, this gives a figure of 95 years; but the true figure is likely to be less because of increased rates of consumption. We must remember also that the conversion of coal to liquid fuels requires energy. Of course, neither coal, nor oil, nor natural gas will disappear entirely, but they will become so expensive that their use as fuels will seem inappropriate, and they will be reserved as starting materials for synthesis.

The date at which the possibility for nuclear energy will end is more controversial and difficult to predict. However, it seems likely that if nuclear reactors are used as an energy source despite their great dangers, finite reserves of uranium and thorium will be exhausted by the end of the 21st century (see Appendix B).

Optimists point to the possibility of using fusion of light elements, such as hydrogen, to generate power. However, although this can be done on a very small scale (and at great expense) in laboratory experiments, the practical generation of energy by means of thermonuclear reactions remains a mirage rather than a realistic prospect on which planners can rely. The reason for this is the enormous temperature required to produce thermonuclear reactions. This temperature is comparable to that existing in the interior of the sun, and it is sufficient to melt any ordinary container. Elaborate "magnetic bottles" have been constructed to contain thermonuclear reactions, and these have been used in successful very small scale experiments. However, despite 50 years of heavily-financed research, there has been absolutely no success in producing thermonuclear energy on a large scale, or at anything remotely approaching commercially competitive prices.

Thus, after the end of the fossil fuel era, our industrial civilization will probably have to rely on renewable sources to supply our energy needs. These sources include hydropower, wind and tidal power, biomass, geothermal energy and solar energy. Let us try to survey how much energy these sources can be expected to produce.

Before the start of the industrial era, human society relied exclusively on renewable energy sources - but can we do so again, with our greatly increased population and greatly increased demands? Will we ultimately be forced to reduce the global population or our per capita use of energy, or both? Let us now try to examine these questions.

8.2 Biomass

Biomass is defined as any energy source based on biological materials produced by photosynthesis - for example wood, sugar beets, rapeseed oil, crop wastes, dung, urban organic wastes, processed sewage, etc. Using biomass for energy does not result in the net emission of CO_2, since the CO_2 released by burning the material had previously been absorbed from the atmosphere during photosynthesis. If the biological material had decayed instead of being burned, it would released the same amount of CO_2 as in the burning process.

The solar constant has the value 1.4 kilowatts/m^2. It represents the amount of solar energy per unit area[1] that reaches the earth, before the sunlight has entered the atmosphere. Because the atmosphere reflects 6% and absorbs 16%, the peak power at sea level is reduced to 1.0 kW/m^2. Clouds also absorb and reflect sunlight. Average cloud cover reduces the energy of sunlight a further 36%. Also, we must take into account the fact that the sun's rays do not fall perpendicularly onto the earth's surface. The angle that they make with the surface depends on the time of day, the season and the latitude.

In Sweden, which lies at a northerly latitude, the solar energy per unit of

[1] The area is assumed to be perpendicular to the sun's rays.

horizontal area is less than for countries nearer the equator. Nevertheless, Göran Persson, the Prime Minister of Sweden, recently announced that his government intends to make the country independent of imported oil by 2020 through a program that includes energy from biomass.

In his thesis, *Biomass in a Sustainable Energy System*, the Swedish researcher Pål Börjesson states that of various crops grown as biomass, the largest energy yields come from short-rotation forests (Salix viminalis, a species of willow) and sugar beet plantations. These have an energy yield of from 160 to 170 GJ_t per hectare-year. (The subscript t means "thermal". Energy in the form of electricity is denoted by the subscript e). One can calculate that this is equivalent to about 0.5 MW_t/km^2, or 0.5 W_t/m^2. Thus, although 1.0 kW/m^2 of solar energy reaches the earth at noon at the equator, the trees growing in northerly Sweden can harvest a day-and-night and seasonal average of only 0.5 Watts of thermal energy per horizontal square meter[2]. Since Sweden's present primary energy use is approximately 0.04 TW_t, it follows that if no other sources of energy were used, a square area of Salix forest 290 kilometers on each side would supply Sweden's present energy needs. This corresponds to an area of 84,000 km^2, about 19% of Sweden's total area[3]. Of course, Sweden's renewable energy program will not rely exclusively on energy crops, but on a mixture of sources, including biomass from municipal and agricultural wastes, hydropower, wind energy and solar energy.

At present, both Sweden and Finland derive about 30% of their electricity from biomass, which is largely in the form of waste from the forestry and paper industries of these two countries.

Despite their northerly location, the countries of Scandinavia have good potentialities for developing biomass as an energy source, since they have small population densities and adequate rainfall. In Denmark, biodiesel oil derived from rapeseed has been used as fuel for experimental buses. Rapeseed fields produce oil at the rate of between 1,000 and 1,300 liters per hectare-crop. The energy yield is 3.2 units of fuel product energy for every unit of fuel energy used to plant the rapeseed, and to harvest and process the oil. After the oil has been pressed from rapeseed, two-thirds of the seed remains as a protein-rich residue which can be fed to cattle.

Miscanthus is a grassy plant found in Asia and Africa. Some forms will also grow in Northern Europe, and it is being considered as an energy crop in the United Kingdom. Miscanthus can produce up to 18 dry tonnes per hectare-year, and it has the great advantage that it can be cultivated using

---

[2]In tropical regions, the rate of biomass production can be more than double this amount.

[3]Additional land area would be needed to supply the energy required for planting, harvesting, transportation and utilization of the wood.

Fig. 8.1 *Rapeseed is a plant belonging to the cabbage or mustard family. The flowers are bright yellow, and the seeds rich in oil.*

ordinary farm machinery. The woody stems are very suitable for burning, since their water content is low (20-30%).

For some southerly countries, honge oil, derived from the plant *Pongamia pinnata* may prove to be a promising source of biomass energy. Studies conducted by Dr. Udishi Shrinivasa at the Indian Institute of Sciences in Bangalore indicate that honge oil can be produced at the cost of

$150 per ton. This price is quite competitive when compared with other potential fuel oils.

Recent studies have also focused on a species of algae that has an oil content of up to 50%. Algae can be grown in desert areas, where cloud cover is minimal. Farm waste and excess CO_2 from factories can be used to speed the growth of the algae.

It is possible that in the future, scientists will be able to create new species of algae that use the sun's energy to generate hydrogen gas. If this proves to be possible, the hydrogen gas may then be used to generate electricity in fuel cells, as will be discussed below in the section on hydrogen technology. Promising research along this line is already in progress at the University of California, Berkeley.

Biogas is defined as the mixture of gases produced by the anaerobic digestion of organic matter. This gas, which is rich in methane (CH_4), is produced in swamps and landfills, and in the treatment of organic wastes from farms and cities. The use of biogas as a fuel is important not only because it is a valuable energy source, but also because methane is a potent greenhouse gas, which should not be allowed to reach the atmosphere. Biogas produced from farm wastes can be used locally on the farm, for cooking and heating, etc. When biogas has been sufficiently cleaned so that it can be distributed in a pipeline, it is known as "renewable natural gas". It may then be distributed in the natural gas grid, or it can be compressed and used in internal combustion engines. Renewable natural gas can also be used in fuel cells, as will be discussed below in the section on Hydrogen Technology.

8.3 Solar energy

Biomass, wind energy, hydropower and wave power derive their energy indirectly from the sun, but in addition, various methods are available for utilizing the power of sunlight directly. These include photovoltaic panels, solar designs in architecture, solar systems for heating water and cooking, concentrating photovoltaic systems, and solar thermal power plants.

Photovoltaic cells and concentrating photovoltaic systems

Solar photovoltaic cells are thin coated wafers of a semiconducting material (usually silicon). The coatings on the two sides are respectively charge donors and charge acceptors. Cells of this type are capable of trapping solar energy and converting it into direct-current electricity. The electricity generated in this way can be used directly (as it is, for example, in pocket calculators) or it can be fed into a general power grid. Alternatively it can be used to split water into hydrogen and oxygen. The gases can then be compressed and stored, or exported for later use in fuel cells. In the future, we may see solar photovoltaic arrays in sun-rich desert areas producing hydrogen as an export product. As their petroleum reserves become exhausted, the countries of the Middle East and Africa may be able to shift to this new technology and still remain energy exporters.

The cost of manufacturing photovoltaic cells is currently falling at the rate of 3%-5% per year. The cost in 2006 was $4.50 per peak Watt. Usually photovoltaic panels are warranted for a life of 20 years, but they are commonly still operational after 30 years or more. The cost of photovoltaic electricity is today 2-5 times the cost of electricity generated from fossil fuels, but photovoltaic costs are falling rapidly, while the costs of fossil fuels are rising equally rapidly.

Concentrating photovoltaic systems are able to lower costs still further by combining silicon solar cells with reflectors that concentrate the sun's rays. The most inexpensive type of concentrating reflector consists of a flat piece of aluminum-covered plastic material bent into a curved shape along one of its dimensions, forming a trough-shaped surface. (Something like this shape results when we hold a piece of paper at the top and bottom with our two hands, allowing the center to sag.) The axis of the reflector can be oriented so that it points towards the North Star. A photovoltaic array placed along the focal line will then receive concentrated sunlight throughout the day.

Photovoltaic efficiency is defined as the ratio of the electrical power produced by a cell to the solar power striking its surface. For commercially available cells today, this ratio is between 9% and 14%. If we assume 5 hours of bright sunlight per day, this means that a photocell in a desert area near to the equator (where 1 kW/m^2 of peak solar power reaches the earth's surface) can produce electrical energy at the average rate of 20-30 W_e/m^2, the average being taken over an entire day and night. Thus the potential power per unit area for photovoltaic systems is far greater than for biomass. However, the mix of renewable energy sources most suitable for a particular country depends on many factors. We saw above that biomass is a promising future source of energy for Sweden, because of Sweden's low population density and high rainfall. By contrast, despite the high initial investment required, photovoltaics are undoubtedly a more promising future energy source for southerly countries with clear skies.

In comparing photovoltaics with biomass, we should be aware of the difference between electrical energy and energy contained in a the chemical bonds of a primary fuel such as wood or rapeseed oil. If Sweden (for example) were to supply all its energy needs from biomass, part of the biomass would have to be burned to generate electricity. The efficiency of energy conversion in electricity generation from fuel is 20%-35%. Of course, in dual use power plants, part of the left-over heat from electrical power generation can be used to heat homes or greenhouses. However, hydropower, wind power and photovoltaics have an advantage in generating electrical power, since they do so directly and without loss, whereas generation of electricity from biomass involves a loss from the inefficiency of the conversion from fuel energy to electrical energy. Thus a rational renewable energy program for Sweden should involve a mixture of biomass for heating and direct fuel use, with hydropower and wind power for generation of electricity. Perhaps photovoltaics will also play a role in Sweden's future electricity generation, despite the country's northerly location and frequently cloudy skies.

The global market for photovoltaics is expanding at the rate of 30% per year. This development is driven by rising energy prices, subsidies to photovoltaics by governments, and the realization of the risks associated with global warming and consequent international commitments to reduce carbon emissions. The rapidly expanding markets have resulted in lowered photovoltaic production costs, and hence further expansion, still lower costs, etc. - a virtuous feedback loop.

Fig. 8.2 *An array of solar parabolic troughs.*

Solar thermal power plants

Solar Parabolic Troughs can be used to heat a fluid, typically oil, in a pipe running along the focal axis. The heated fluid can then be used to generate electrical power. The liquid that is heated in this way need not be oil. In a solar thermal power plant in California, reflectors move in a manner that follows the sun's position and they concentrate solar energy onto a tower, where molten salt is heated to a temperature of 1050 degrees F (566 degrees C). The molten salt stores the heat, so that electricity can be generated even when the sun is not shining. The California plant, now in a three-year operating and testing phase, generates 10 MW_e.

Fig. 8.3 *Solar Two, an experimental solar thermal power plant in California's Mojave Desert. Arrays of heliostatic reflectors concentrate the sun's rays onto molten salt in the tower. The plant produces 10 MW_e.*

Solar designs in architecture

At present, the average global rate of use of primary energy is roughly 2 kW_t per person. In North America, the rate is 12 kW_t per capita, while in Europe, the figure is 6 kW_t. In Bangladesh, it is only 0.2 kW_t. This wide variation implies that considerable energy savings are possible, through changes in lifestyle, and through energy efficiency.

Important energy savings can be achieved through solar design in architecture. For example, insulation can be improved in walls, and insulating shutters can be closed at night.

In double envelope construction, a weatherproof shell surrounds the inner house. Between the outer shell and the house, sun-heated air circulates.

Fig. 8.4 *A solar cooker.*

A less extreme example of this principle is the construction of south-facing conservatories. The sun-heated air in the conservatories acts as a thermal buffer, and reduces heat loss from the house.

Solar design aims at making houses cool in the summer and warm in the winter. Awnings can be spread out in the summer to shade windows, and rolled together in the winter to allow sunshine to enter the house. Alternatively, deciduous trees can be planted in front of south-facing windows. During the summer, the leaves of the trees shade the windows, while in the winter, the leaves fall, allowing the sun to enter.

During daylight hours, houses can be illuminated by fiber optic light pipes, connected to a parabolic collector on the roof. The roof can also contain arrays of solar photovoltaic cells and solar water heaters.

Houses can be heated in the winter by heat pumps connected to a deeply buried network of pipes. Heat pumps function in much the same way as

Fig. 8.5 *Solar thermal panels at Ulsted in Denmark.*

refrigerators or air conditioners. When they are used to warm houses in the winter, a volatile liquid such as ammonia is evaporated underground, where the temperature is relatively constant, not changing much between summer and winter. In the evaporation process, heat is absorbed from the ground. The gas is then compressed and reliquefied within the house, and in this process, it releases the heat that was absorbed underground. Electricity is of course required to drive a heat pump, but far less electrical power is needed to do this than would be required to heat the house directly.

In general, solar design of houses and other buildings requires an initial investment, but over time, the investment is amply repaid through energy savings.

Solar systems for heating water and cooking

Solar heat collectors are are already in common use to supply hot water for families or to heat swimming pools. A common form of the solar heat collector consists of a flat, blackened heat-collecting plate to which tubes containing the fluid to be heated are connected. The plate is insulated from the atmosphere by a layer of air (in some cases a partial vacuum) above which there is a sheet of glass. Water flowing through the tubes is collected in a tank whenever it is hotter than the water already there. In cases where there is a danger of freezing, the heated fluid may contain antifreeze, and it may then exchange heat with water in the collection tank. Systems of this kind can function even in climates as unfavorable as that of Northern Europe, although during winter months they must be supplemented by conventional water-heaters.

In the developing countries, wood is often used for cooking, and the result is sometimes deforestation, soil erosion and desertification. In order to supply an alternative, many designs for solar cooking have been developed. Often the designs are very simple, and many are both easy and inexpensive to build, the starting materials being aluminum foil and cardboard boxes.

8.4 Wind energy

Wind parks in favorable locations, using modern wind turbines, are able to generate 10 MW_e/km^2 or 10 W_e/m^2. Often wind farms are placed in offshore locations. When they are on land, the area between the turbines can be utilized for other purposes, for example for pasturage. For a country like Denmark, with good wind potential but cloudy skies, wind turbines can be expected to play a more important future role than photovoltaics. Denmark is already a world leader both in manufacturing and in using wind turbines. Today, 23% of all electricity used in Denmark is generated by wind power, and the export of wind turbines makes a major contribution to the Danish economy.

Globally, less than 1% of all electricity generated comes from wind power. This corresponds to 59 GW_e or 0.059 TW_e. However, the use of wind power is currently growing at the rate of 38% per year. In the United States, it is the fastest-growing form of electricity generation.

The location of wind parks is important, since the energy obtainable from wind is proportional to the cube of the wind velocity. We can understand this cubic relationship by remembering that the kinetic energy of a moving object is proportional to the square of its velocity multiplied by the mass. Since the mass of air moving past a wind turbine is proportional to the wind velocity, the result is the cubic relationship just mentioned.

Before the decision is made to locate a wind park in a particular place, the wind velocity is usually carefully measured and recorded over an entire year. For locations on land, mountain passes are often very favorable locations, since wind velocities increase with altitude, and since the wind is concentrated in the passes by the mountain barrier. Other favorable locations include shorelines and offshore locations on sand bars. This is because

Fig. 8.6 *Installation of a wind turbine.*

onshore winds result when warm air rising from land heated by the sun is replaced by cool marine air. Depending on the season, the situation may be reversed at night, and an offshore wind may be produced if the water is warmer than the land.

The cost of wind-generated electrical power is currently about 5 US cents per kilowatt hour, i.e., lower than the cost of electricity generated by burning fossil fuels.

The "energy payback ratio" of a power installation is defined as the ratio of the energy produced by the installation over its lifetime, divided by the energy required to manufacture, construct, operate and decommission the installation. For wind turbines, this ratio is 17-39, compared with 11 for coal-burning plants. The construction energy of a wind turbine is usually paid back within three months.

Besides the propeller-like design for wind turbines there are also designs where the rotors turn about a vertical shaft. One such design was patented in 1927 by the French aeronautical engineer Georges Jean Marie Darrieus. The blades of a Darrieus wind turbine are airfoils similar to the wings of an aircraft. As the rotor turns in the wind, the stream of air striking the airfoils produces a force similar to the "lift" of an airplane wing. This force pushes the rotor in the direction that it is already moving. The Darrieus design has some advantages over conventional wind turbine design, since the generator can be placed at the bottom of the vertical shaft, where it may be more easily serviced. Furthermore, the vertical shaft can be lighter than the shaft needed to support a conventional wind turbine.

One problem with wind power is that it comes intermittently, and demand for electrical power does not necessarily come at times when the wind is blowing most strongly. To deal with the problem of intermittency, wind power can be combined with other electrical power sources in a grid. Alternatively, the energy generated can be stored, for example by pumped hydroelectric storage or by using hydrogen technology, as will be discussed below.

Bird lovers complain that birds are sometimes killed by rotor blades. This is true, but the number killed is small. For example, in the United States, about 70,000 birds per year are killed by turbines, but this must be compared with 57 million birds killed by automobiles and 97.5 million killed by collisions with plate glass.

The aesthetic aspects of wind turbines also come into the debate. Perhaps in the future, as wind power becomes more and more a necessity and less a matter of choice, this will be seen as a "luxury argument".

8.5 Hydroelectric power

At present 20% of the world's electricity comes from hydroelectric power. In the developed countries, the potential for increasing this percentage is small, because most of the suitable sites for dams are already in use. Mountainous regions of course have the greatest potential for hydroelectric power, and this correlates well with the fact that virtually all of the electricity generated in Norway comes from hydro, while in Iceland and Austria the figures are respectively 83% and 67%. Among the large hydroelectric power stations now in use are the La Grande complex in Canada (16 GW_e) and the Itapú station on the border between Brazil and Paraguay (14 GW_e). The Three

Fig. 8.7 *Aswan High Dam, Egypt, viewed from space.*

Gorges Dam under construction in China is planned to produce 18.2 GW_e by 2009.

Even in regions where the percentage of hydro in electricity generation is not so high, it plays an important role because hydropower can be used selectively at moments of peak demand. Pumping of water into reservoirs can also be used to store energy.

The creation of lakes behind new dams in developing countries often involves problems, for example relocation of people living on land that will be covered by water, and loss of the land for other purposes[4]. However the energy gain per unit area of lake can be very large - over 100 W_e/m^2. Fish ladders can be used to enable fish to reach their spawning grounds above dams. In addition to generating electrical power, dams often play useful roles in flood control and irrigation.

At present, hydroelectric power is used in energy-intensive industrial processes, such as the production of aluminum. However, as the global

[4]Over a million people were displaced by the construction of the Three Gorges Dam in China, and many sites of cultural value were lost.

Table 8.1: Technical potential and utilization of hydropower. (Data from World Energy Council, 2003.)

Region	Technical potential	Annual output	Percent used
Asia	0.5814 TW_e	0.0653 TW_e	11%
S. America	0.3187 TW_e	0.0579 TW_e	18%
Europe	0.3089 TW_e	0.0832 TW_e	27%
Africa	0.2155 TW_e	0.0091 TW_e	4%
N. America	0.1904 TW_e	0.0759 TW_e	40%
Oceania	0.0265 TW_e	0.0046 TW_e	17%
World	1.6414 TW_e	0.2960 TW_e	18%

energy crisis becomes more severe, we can expect that metals derived from electrolysis, such as aluminum and magnesium, will be very largely replaced by other materials, because the world will no longer be able to afford the energy needed to produce them.

8.6 Energy from the ocean

Tidal power

The twice-daily flow of the tides can be harnessed to produce electrical power. Ultimately tidal energy comes from the rotation of the earth and its interaction with the moon's gravitational field. The earth's rotation is very gradually slowing because of tidal friction, and the moon is gradually receding from the earth, but this process will take such an extremely long time that tidal energy can be thought of as renewable.

There are two basic methods for harnessing tidal power. One can build barriers that create level differences between two bodies of water, and derive hydroelectric power from the head of water thus created. Alternatively it is possible to place the blades of turbines in a tidal stream. The blades are then turned by the tidal current in much the same way that the blades of a wind turbine are turned by currents of air.

There are plans for using the second method on an extremely large scale in Cook Strait, near New Zealand. A company founded by David Beach and Chris Bathurst plans to anchor 7,000 turbines to the sea floor of Cook Strait in such a way that they will float 40 meters below the surface. Beach and Bathurst say that in this position, the turbines will be safe from the effects of earthquakes and storms. They plan to have the first turbine in place by 2008. The tidal flow through Cook Strait is so great that the scheme could supply all of New Zealand's electricity if the project is completed on the scale visualized by its founders.

Choosing the proper location for tidal power stations is important, since the height of tides depends on the configuration of the land. For example, tides of 17 meters occur in the Bay of Fundy, at the upper end of the Gulf of Maine, between New Brunswick and Nova Scotia. Here tidal waves are funneled into the bay, creating a resonance that results in the world's greatest level difference between high an low tides. An 18 MW_e dam-type tidal power generation station already exists at Annapolis River, Nova Scotia, and there are proposals to increase the use of tidal power in the Bay of Fundy. Some proposals involve turbines in the tidal stream, similar to those proposed for use in the Cook Strait.

In the future, favorable locations for tidal power may be exploited to their full potentialities, even thought the output of electrical energy exceeds local needs. The excess energy can be stored in the form of hydrogen (see below) and exported to regions deficient in renewable energy resources.

Wave energy

At present, the utilization of wave energy is in an experimental stage. In Portugal, there are plans for a wave farm using the Pelamis Wave Energy Converter. The Pelamis is a long floating tube with two or more rigid sections joined by hinges. The tube is tethered with its axis in the direction of wave propagation. The bending between sections resulting from passing waves is utilized to drive high pressure oil through hydraulic motors coupled to electrical generators. Each wave farm in the Portuguese project is planned to use three Pelamis converters, each capable of producing 750 kW_e. Thus the total output of each wave farm will be 2.25 MW_e.

Another experimental wave energy converter is Salter's Duck, invented in the 1970's by Prof. Stephen Salter of the University of Edinburgh, but still being developed and improved. Like the Pelamis, the Duck is also cylindrical in shape, but the axis of the cylinder is parallel to the wave front, i.e. perpendicular to the direction of wave motion. A floating cam, attached to the cylinder, rises and falls as a wave passes, driving hydraulic motors within the cylinder. Salter's Duck is capable of using as much as 65% of the wave's energy.

The energy potentially available from waves is very large, amounting to as much as 100 kilowatts per meter of wave front in the best locations.

Ocean thermal energy conversion

In tropical regions, the temperature of water at the ocean floor is much colder than water at the surface. In ocean thermal energy conversion, cold water is brought to the surface from depths as great as 1 km, and a heat engine is run between deep sea water at a very low temperature and surface water at a much higher temperature.

According to thermodynamics, the maximum efficiency of a heat engine operating between a cold reservoir at the absolute temperature T_C and a hot reservoir at the absolute temperature T_H is given by $1-T_C/T_H$. In order to convert temperature on the centigrade scale to absolute temperature (degrees Kelvin) one must add 273 degrees. Thus the maximum efficiency of a heat engine operating between water at the temperature of 25 degrees C and water at 5 degrees C is $1-(5+273)/(25+273)=0.067 = 6.7\%$. The efficiency of heat engines is always less than the theoretical maximum because of various losses, such as the loss due to friction. The actual overall efficiencies of existing ocean thermal energy conversion (OTEC) stations are typically 1-3%. On the other hand, the amount of energy potentially available from differences between surface and bottom ocean temperatures is extremely large.

Since 1974, OTEC research has been conducted by the United States at the Natural Energy Laboratory of Hawaii. The Japanese government also supports OTEC research, and India has established a 1 MW_e OTEC power station floating in the ocean near to Tamil Nadu.

Methane clatherates

A clatherate is a crystal formed primarily of water, which forms cagelike structures around atoms or molecules of some other substance. In methane clatherates (sometimes called methane hydrate crystals) the molecule methane (CH_3) is surrounded by a cage of water molecules. Methane clatherates can be stable at temperatures of up to 18 degrees C, depending on the pressure.

Methane clatherate crystals are present in very large amounts in various locations on the ocean floor. At present, it is estimated that globally the amount is 1-5$\times10^{15}$ m^3. This corresponds to 500-2500 gigatonnes of carbon (GtC). An additional 400 GtC of methane clatherate crystals are thought to be present in the permafrost of the Arctic. These figures can be compared with the 5000 GtC estimated to be the amount of all other fossil fuels. Whether or not harvesting of methane from ocean-floor clatherates can be performed in an economically viable way remains to be seen.

From the isotope ratios in the methane present it can be deduced that much of the total amount of methane clatherate has been produced by anaerobic bacterial-aided decay of organic matter. The speed of this process affects our decision on whether to regard methane clatherates as fossil fuels or whether we should view them as renewable on a human timescale, i.e., as a form of biogas.

The large amounts of methane trapped in clatherates can be regarded as a threat as well as a promise, since methane is a potent greenhouse gas. Many experts fear that as frozen Arctic tundra is melted by global warming, methane will be released into the atmosphere, thus further accelerating the process of climate change. The same can be said for methane clatherates ocean floors. If sea temperatures rise substantially, the crystals may become unstable, and the methane may be released into the atmosphere.

8.7 Geothermal energy

The ultimate source of geothermal energy is the decay of radioactive nuclei in the interior of the earth. Because of the heat produced by this radioactive decay, the temperature of the earth's core is 4300 degrees C. The inner core is composed of solid iron, while the outer core consists of molten iron and sulfur compounds. Above the core is the mantle, which consists of a viscous liquid containing compounds of magnesium, iron, aluminum, silicon and oxygen. The temperature of the mantle gradually decreases from 3700 degrees C near the core to 1000 degrees C near the crust. The crust of the earth consists of relatively light solid rocks and it varies in thickness from 5 to 70 km.

Fig. 8.8 *A geothermal power plant near Grindavik, Iceland.*

The outward flow of heat from radioactive decay produces convection currents in the interior of the earth. These convection currents, interacting with the earth's rotation, produce patterns of flow similar to the trade winds of the atmosphere. One result of the currents of molten conducting material in the interior of the earth is the earth's magnetic field. The crust is divided into large sections called "tectonic plates", and the currents of molten material in the interior of the earth also drag the plates into collision with each other. At the boundaries, where the plates collide or split apart, volcanic activity occurs. Volcanic regions near the tectonic plate boundaries are the best sites for collection of geothermal energy.

The entire Pacific Ocean is ringed by regions of volcanic and earthquake activity, the so-called Ring of Fire. This ring extends from Tierra del Fuego at the southernmost tip of South America, northward along the western coasts of both South America and North America to Alaska. The ring then crosses the Pacific at the line formed by the Alutian Islands, and it reaches the Kamchatka Peninsula in Russia. From there it extends southward along the Kuril Island chain and across Japan to the Philippine Islands, Indonesia and New Zealand. Many of the islands of the Pacific are volcanic in nature. Another important region of volcanic activity extends northward along the Rift Valley of Africa to Turkey, Greece and Italy. In the Central Atlantic region, two tectonic plates are splitting apart, thus producing the volcanic activity of Iceland. All of these regions are very favorable for the collection of geothermal power.

The average rate at which the energy created by radioactive decay in the interior of the earth is transported to the surface is 0.06 W_t/m^2. However, in volcanic regions near the boundaries of tectonic plates, the rate at which the energy is conducted to the surface is much higher - typically 0.3 W_t/m^2. If we insert these figures into the thermal conductivity law

$$q = K_T \frac{\Delta T}{z}$$

we can obtain an understanding of the types of geothermal resources available throughout the world. In the thermal conductivity equation, q is the power conducted per unit area, while K_T is the thermal conductivity of the material through which the energy is passing. For sandstones, limestones and most crystalline rocks, thermal conductivities are in the range 2.5-3.5 $W_t/(m\ ^oC)$. Inserting these values into the thermal conductivity equation, we find that in regions near tectonic plate boundaries we can reach temperatures of 200 ^oC by drilling only 2 kilometers into rocks of the types named above. If the strata at that depth contain water, it will be in the form of highly-compressed steam. Such a geothermal resource is called a *high-enthalpy* resource[5].

In addition to high-enthalpy geothermal resources there are *low-enthalpy* resources in nonvolcanic regions of the world, especially in basins covered by sedimentary rocks. Clays and shales have a low thermal conductivity, typically 1-2 $W_t/(m\ ^oC)$. When we combine these figures with the global average geothermal power transmission, $q = 0.06\ W_t/m^2$, the thermal conduction equation tells us that $\Delta T/z = 0.04\ ^oC/m$. In such a region the geothermal resources may not be suitable for the generation of electrical power, but nevertheless adequate for heating buildings. The Creil district

[5]Enthalpy $\equiv H \equiv U + PV$ is a thermodynamic quantity that takes into account not only the internal energy U of a gas, but also energy PV that may be obtained by allowing it to expand.

heating scheme north of Paris is an example of a project where geothermal energy from a low enthalpy resource is used for heating buildings.

The total quantity of geothermal electrical power produced in the world today is 8 GW_e, with an additional 16 GW_t used for heating houses and buildings. In the United States alone, 2.7 GW_e are derived from geothermal sources. In some countries, for example Iceland and Canada, geothermal energy is used both for electrical power generation and for heating houses.

There are three methods for obtaining geothermal power in common use today: Deep wells may yield dry steam, which can be used directly to drive turbines. Alternatively water so hot that it boils when brought to the surface may be pumped from deep wells in volcanic regions. The steam is then used to drive turbines. Finally, if the water from geothermal wells is less hot, it may be used in binary plants, where its heat is exchanged with an organic fluid which then boils. In this last method, the organic vapor drives the turbines. In all three methods, water is pumped back into the wells to be reheated. The largest dry steam field in the world is The Geysers, 145 kilometers north of San Francisco, which produces 1,000 MW_e.

There is a fourth method of obtaining geothermal energy, in which water is pumped down from the surface and is heated by hot dry rocks. In order to obtain a sufficiently large area for heat exchange the fissure systems in the rocks must be augmented, for example by pumping water down at high pressures several hundred meters away from the collection well. The European Union has established an experimental station at Soultz-sous-Forets in the Upper Rhine to explore this technique. The experiments performed at Soultz will determine whether the "hot dry rock" method can be made economically viable. If so, it can potentially offer the world a very important source of renewable energy.

The molten lava of volcanoes also offers a potential source of geothermal energy that may become available in the future, but at present, no technology has been developed that is capable of using it.

8.8 Hydrogen technologies

Electrolysis of water

When water containing a little acid is placed in a container with two electrodes and subjected to an external direct current voltage greater than 1.23 Volts, bubbles of hydrogen gas form at one electrode (the cathode), while bubbles of oxygen gas form at the other electrode (the anode). At the cathode, the half-reaction

$$2H_2O(l) \rightarrow O_2(g) + 4H^+(aq) + 4e^- \qquad E^0 = -1.23\ Volts$$

takes place, while at the anode, the half-reaction

$$4H^+(aq) + 4e^- \rightarrow 2H_2(g) \qquad E^0 = 0$$

occurs.

Half-reactions differ from ordinary chemical reactions in containing electrons either as reactants or as products. In electrochemical reactions, such as the electrolysis of water, these electrons are either supplied or removed by the external circuit. When the two half-reactions are added together, we obtain the total reaction:

$$2H_2O(l) \rightarrow O_2(g) + 2H_2(g) \qquad E^0 = -1.23\ Volts$$

Notice that $4H^+$ and $4e^-$ cancel out when the two half-reactions are added. The total reaction does not occur spontaneously (as is discussed in Appendix A), but it can be driven by an external potential E, provided that the magnitude of E is greater than 1.23 volts.

When this experiment is performed in the laboratory, platinum is often used for the electrodes, but electrolysis of water can also be performed using electrodes made of graphite.

Electrolysis of water to produce hydrogen gas has been proposed as a method for energy storage in a future renewable energy system. For example, it might be used to store energy generated by photovoltaics in desert areas of the world. Compressed hydrogen gas could then be transported to other regions and used in fuel cells. Electrolysis of water and storage of hydrogen could also be used to solve the problem of intermittency associated with wind energy or solar energy.

Hydrogen fuel cells

Fuel cells allow us to convert the energy of chemical reactions directly into electrical power. In hydrogen fuel cells, for example, the exact reverse of the electrolysis of water takes place. Hydrogen reacts with oxygen, and produces electricity and water, the reaction being

$$O_2(g) + 2H_2(g) \rightarrow 2H_2O(l) \qquad E^0 = 1.23\ Volts$$

The arrangement of the a hydrogen fuel cell is such that the hydrogen cannot react directly with the oxygen, releasing heat. Instead, two half reactions take place, one at each electrode, as was just mentioned in connection with the electrolysis of water. In a hydrogen fuel cell, hydrogen gas produces electrons and hydrogen H^+ ions at one of the electrodes.

$$2H_2(g) \rightarrow 4H^+(aq) + 4e^- \qquad E^0 = 0$$

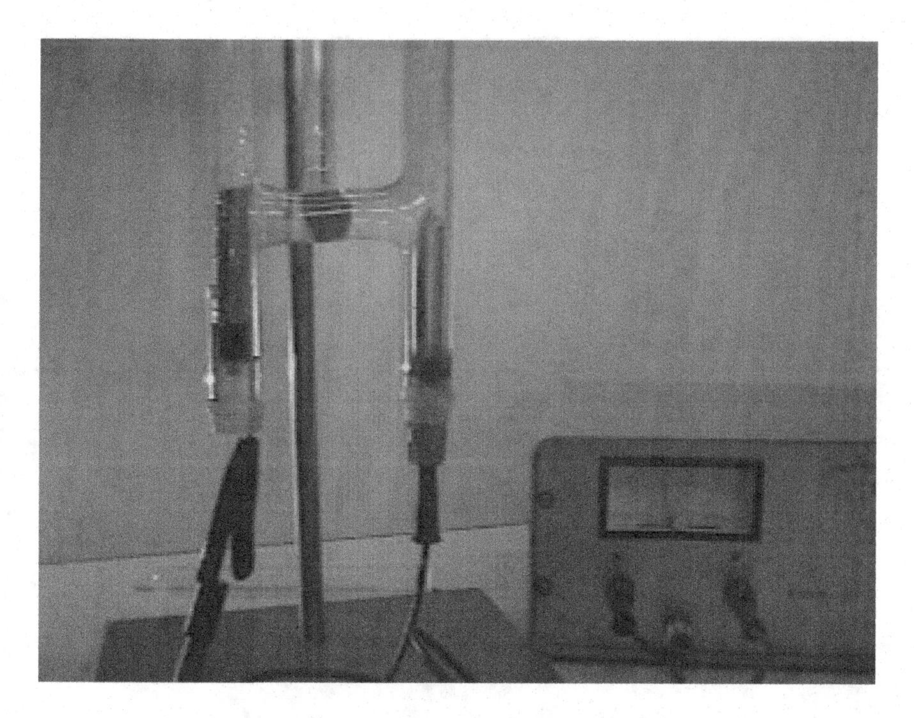

Fig. 8.9 *Electrolysis of water.*

The electrons flow through the external circuit to the oxygen electrode, while the hydrogen ions complete the circuit by flowing through the interior of the cell (from which the hydrogen and oxygen molecules are excluded by semipermeable membranes) to the oxygen electrode. Here the electrons react with oxygen molecules and H^+ ions to form water.

$$O_2(g) + 4H^+(aq) + 4e^- \rightarrow 2H_2O(l) \qquad E^0 = 1.23\ Volts$$

In this process, a large part of the chemical energy of the reaction becomes available as electrical power.

We can recall that the theoretical maximum efficiency of a heat engine operating between a cold reservoir at temperature T_C and a hot reservoir at T_H is $1\text{-}T_C/T_H$, where the temperatures are expressed on the Kelvin scale. Since fuel cells are not heat engines, their theoretical maximum efficiency is not limited in this way. Thus it can be much more efficient to generate electricity by reacting hydrogen and oxygen in a fuel cell than it would be to burn the hydrogen in a heat engine and then use the power of the engine to drive a generator.

Hydrogen technologies are still at an experimental stage. Furthermore,

Fig. 8.10 *A methanol fuel cell.*

they do not offer us a source of renewable energy, but only means for storage, transportation and utilization of energy derived from other sources. Nevertheless, it seems likely that hydrogen technologies will have great importance in the future.

8.9 Some concluding remarks

It can be seen from our discussion of renewable energy technologies that they can potentially offer at least a partial replacement for the fossil fuels on which the world is now dependent. All forms of renewable energy should be developed simultaneously, since all will be needed. Energy conservation and changes of lifestyle will also be necessary.

It seems likely that biomass, photovoltaics, solar thermal power, wind power and wave power will become the major energy sources of the future. In addition, hydropower is extremely helpful in overcoming the problem of intermittency, while other forms of renewable energy may have great advantages in certain locations.

The transition to renewable energy will require wholehearted governmental commitment, tax changes, and a considerable investment in research. At present nuclear energy, nuclear research and the oil industry all receive enormous governmental support. It is vital that this support should go instead to renewable energy technologies.

The time factor is also important. The Hubbert peak for oil will occur in a decade, and the peak for natural gas in two decades. After that, the outlook for the future is that petroleum and natural gas will become more and more expensive - finally so expensive that they will not be used as fuels. To minimize the shock of these events, serious work on substitutes must begin immediately, and on a large scale. At present the development of renewable energy is proceeding so slowly that if the trend is not corrected, we can anticipate a period of energy scarcity and economic trauma.

The transition to renewable energy will involve rededication of much land from agriculture to energy generation. This will be easiest in countries where the population density is low, and difficult in countries that already have problems in feeding their people. In the era beyond fossil fuels, the optimum global population will perhaps be less than the present population of the world. The optimum population is one that can be supported in comfort and dignity, and with respect for the environment.

Suggestions for further reading

(1) G. Boyle (editor), *Renewable Energy: Power for a Sustainable Future, Second Edition*, Oxford University Press, (2004).
(2) G. Boyle, B. Everett and J. Ramage (editors), *Energy Systems and Sustainability*, Oxford University Press, (2003).
(3) United Nations Development Programme, *World Energy Assessment*, United Nations, New York, (2002).
(4) P. Smith et al., *Meeting Europe's Climate Change Commitments: Quantitative Estimates of the Potential for Carbon Mitigation by Agriculture*, Global Change Biology, **6**, 525-39, (2000).
(5) R.E.H. Sims, *The Brilliance of Energy: In Business and in Practice*, James and James, London, (2002).
(6) Friends of the Earth, *Energy Without End*, FOE, London, (1991).
(7) E.A. Alma and E. Neala, *Energy Viability of Photovoltaic Systems*, Energy Policy, **28**, 999-1010, (2000).
(8) Department of Trade and Industry, *Developments of Solar Photovoltaics in Japan*, Global Watch Mission Report, November, (2003).
(9) European Photovoltaics Industry Association, *Solar Generation: Solar Electricity for Over 1 Billion People and 2 Million Jobs by 2020*, EPA, published in association with Greenpeace (see www.cleanenergynow.org), (2001).
(10) B. Sørensen, *Renewable Energy, Second Edition*, Academic Press, (2000).

(11) K. Illum, *A Viable Energy Strategy for the Nordic Countries*, Greenpeace Nordic, (2006).
(12) K. Illum, *SESAME: The Sustainable Energy Systems Analysis Model*, Along University Press, Denmark, (1995).
(13) G. Sinden, *Wind Power and the UK Wind Resource*, Environmental Change Institute, University of Oxford, (2005).
(14) P. Börjesson, *Energy Analysis of Biomass Production and Transportation*, Biomass and Energy, 11, 305-318, (1996).
(15) P. Börjesson, *Emissions of CO2 from Biomass Production and Transportation*, Energy Conversion Management, 37, 1235-1240, (1995).
(16) P. Börjesson and L. Gustav's, *Regional Production and Utilization of Biomass in Sweden*, Energy - The International Journal, 21, 747-764, (1996).
(17) D. Duodena and C. Galvin, *Renewable Energy*, W.W. Norton, New York, (1983).
(18) G. Foley, *The Energy Question*, Penguin Books Ltd., (1976).
(19) D. Hayes, *The Solar Energy Timetable*, Worldwatch Paper 19, Worldwatch Institute, Washington D.C., (1978).
(20) J. Holdren and P. Herrera, *Energy*, Sierra Club Books, New York, (1971).
(21) L. Rosen and R.Glasser (eds.), *Climate Change and Energy Policy*, Los Alamos National Laboratory, AIP, New York, (1992).
(22) A.B. Lovins, *Soft Energy Paths*, Ballinger, Cambridge, (1977).
(23) National Academy of Sciences, *Energy and Climate*, NAS, Washington D.C., (1977).
(24) LTI-Research Group, ed., *Long-Term Integration of Renewable Energy Sources into the European Energy System*, Physica Verlag, (1998).
(25) P.H. Ableson, *Renewable Liquid Fuels*, Science, **268**, 5213, (1995).
(26) W.H. Avery and C. Wu, *Renewable Energy From the Ocean. A Guide to OTEC*, Oxford University Press, (1994).
(27) H.M. Browenstein et al., *Biomass Energy Systems and the Environment*, Pergamon Press, New York, (1981).
(28) M. Brower, *Cool Energy. Renewable Solutions to Environmental Problems*, MIT Press, Cambridge, Massachusetts, (1994).
(29) L.R. Brown, *Food or Fuel: New Competition for the World's Cropland*, Worldwatch Paper 35, Worldwatch Institute,Washington D.C., (1980).
(30) C.H. Deutsch, *As Oil Prices Rise, the Hydrogen Car is Looking Better*, New York Times, August 26, (1990).
(31) W.L. Driscoll, *Fill 'Er Up with Biomass Derivatives*, Technology Review, August/September, 74-76, (1993).
(32) J. Dunkerley et al., *Energy Strategies for Developing Nations*, Resources for the Future, Washington D.C., (1981).

(33) C. Flavin and N. Lenssen, *Beyond the Petroleum Age: Designing a Solar Economy*, Worldwatch Paper 100, Worldwatch Institute, Washington D.C., (1990).
(34) C. Flavin and N. Lenssen, *The Power Surge. Guide to the Coming Energy Revolution*, W.W. Norton, New York, (1994).
(35) J. Gever et al., *Beyond Oil. The Threat to Food and Fuel in the Coming Decades*, Ballinger, Cambridge Massachusetts, (1986).
(36) E.A. Hiller and B.A. Stout, *Biomass Energy. A Monograph*, Texas A&M University Press, College Station, Texas, (1985).
(37) T.B. Johansson et al., (editors), *Renewable Energy: Sources for Fuels and Electricity*, Island Press, Washington D.C., (1982).
(38) D. Knott, *Hydrogen: The Fuel of the Future?*, Oil and Gas Journal, May, 26, (1994).
(39) D.K. McDaniels, *The Sun: Our Future Energy Source* Second Edition, Krieger, Malabar Florida, (1994).
(40) D. Pimentel, *Renewable Energy: Economic and Environmental Issues*, BioScience, **44**, 536-547, (1994).
(41) D. Ross, *Power From the Waves*, Oxford University Press, (1995).
(42) S. Sanchez, *Movement is in the Air As Texas Taps the Wind*, USA Today, November, (1995).
(43) C.C. Swan, *Suncell. Energy, Economics and Photovoltaics*, Sierra Club Books, San Francisco, (1986).
(44) I. Wickelgren, *Sunup at Last for Solar?*, Business Week, July 24, 84,86, (1995).
(45) K. Zweibel, *Thin-film Photovoltaic Cells*, American Scientist, **81**, 362-369, (1993).
(46) R. Golob and E. Brus, *The Almanac of Renewable Energy. The Complete Guide to Emerging Energy Technologies*, Henry Holt and Company, New York, (1993).
(47) N. Yoneda and S. Ito, *Study of Energy Storage for Long Term Using Chemical Reactions*, 3rd International Solar Forum, Hamburg, Germany, June 24-27, (1980).
(48) R. Curtis, *Earth Energy in the UK*, in *Proc. International Geothermal Days 'Germany 2001', conference*, Bad Urach, Available in PDF format on www.uni-giessen.de, (2001).
(49) R. Harrison, N.D. Mortimer and O.B. Smarason, *Geothermal Heating: A Handbook of Engineering Economics*, Pergamon Press, (1990).
(50) G.W. Huttrer, *The Status of World Geothermal Power Generation 1995-2000*, in *WGC 2000*, (2000).
(51) International Geothermal Association, *Performance Indicators for Geothermal Power Plant*, IGA News, **45**, July-September, (2001).

(52) J.W. Lund and D.H. Freestone, *Worldwide Direct Use of Geothermal Energy 2000*, in *WGC 2000*, (2000).
(53) H.C.H. Armstead and J.W. Tester, *Heat Mining*, Chapman and Hall, (1987).
(54) T.B. Johansson et al. (editors), *Electricity - Efficient End Use*, Lund University Press, (1989).
(55) S. Krohn, *Wind Energy Policy in Denmark, Status 2002*, Danish Windpower Association, (2002), www.windpower.dk.
(56) Imperial College London, *Assessment of Technological Options to Address Climate Change*, ICCEPT, (2002), www.iccept.ic.ac.uk.
(57) J. Guldemberg, (editor), *World Energy Assessment: Energy and the Challenge of Sustainability*, United Nations Development Programme, New York, (2000).
(58) European Commission, *Green Paper - Towards a European Strategy for the Security of Energy Supply*, COM(2000) final, (2000).
(59) P. Dal and H.S. Jensen, *Energy Efficiency in Denmark*, Danish Energy Ministry, (2000).
(60) A.C. Baker, *Tidal Power*, Peter Peregrinus, (1991).
(61) A.C. Baker, *Tidal Power*, Energy Policy, **19**, 792-7, (1991).
(62) Department of Energy, *Tidal Power from the Severn Estuary, Volume I*, Energy Paper 46 (The Bondi Report), HMSO, (1987).
(63) House of Commons Select Committee on Energy, *Renewable Energy*, Fourth Report Session 1991-1992, HMSO, (1992).
(64) American Wind Energy Association, *Global Wind Energy Market Report*, AWEA, (2003).
(65) M. Andersen, *Current Status of Wind Farms in the UK*, Renewable Energy Systems, (1992).
(66) Border Wind, *Offshore Wind Energy: Building a New Industry for Britain*, Greenpeace, (1998).
(67) J. Beurkens and P.H. Jensen, *Economics of Wind Energy, Prospects and Directions*, Renewable Energy World, July-Aug, (2001).
(68) British Wind Energy Association, *Best Practice Guidelines for Wind Energy Development*, BWEA, (1994).
(69) British Wind Energy Association, *Planning Progress*, BWEA, (2003) (www.britishwindenergy.co.uk).
(70) Department of Trade and Industry, *Future Offshore: A Strategic Framework for the Offshore Wind Industry*, DTI, November (2002).
(71) European Wind Energy Association, *Time for Action: Wind Energy in Europe*, EWEA, (1991).
(72) European Wind Energy Association, *Wind Force 12*, EWEA, (2002).
(73) T. Burton et al., *Wind Energy Handbook*, Wiley, (2001).
(74) J.F. Manwell et al., *Wind Energy Explained*, Wiley, (2002).

Chapter 9

ECONOMICS WITHOUT GROWTH

9.1 The transition from growth to a steady state - minimizing the trauma

According to Adam Smith, the free market is the dynamo of economic growth. The true entrepreneur does not indulge in luxuries for himself and his family, but reinvests his profits, with the result that his business or factory grows larger, producing still more profits, which he again reinvests, and so on. This is indeed the formula for exponential economic growth.

Economists (with a few notable exceptions) have long behaved as though growth were synonymous with economic health. If the gross national product of a country increases steadily by 4% per year, most economists express approval and say that the economy is healthy. If the economy could be made to grow still faster (they maintain), it would be still more healthy. If the growth rate should fall, economic illness would be diagnosed. However, the basic idea of Malthus is applicable to exponential increase of any kind. It is obvious that on a finite Earth, neither population growth nor economic growth can continue indefinitely.

A "healthy" economic growth rate of 4% per year corresponds to an increase by a factor of 50 in a century, by a factor of 2500 in two centuries, and by a factor of 125,000 in three centuries. No one can maintain that this type of growth is sustainable except by refusing to look more than a short distance into the future.

But *why* do most economists cling so stubbornly and blindly to the concept of growth? Why do they refuse to look more than a few years into the future? We can perhaps understand this strange self-imposed myopia by remembering some of David Ricardo's ideas: One of his most important contributions to economic theory was his analysis of rents. Ricardo considered the effects of economic expansion; and he concluded that as population increased, marginally fertile land would be forced into cultivation.

The price of grain would be determined by the cost of growing it on inferior land; and the owners of better land would be able to pocket a progressively larger profit as worse and worse land was forced into use by the demands of a growing population. Ricardo's analysis of rents for agricultural land has various generalizations; for example, a growing population also puts pressure on land used for building cities, and profits can be gained by holding such land, or through the ownership of houses in growing cities. In general, in a growing economy, investments are likely to be rewarded. In a stationary or contracting economy, the stock market may crash.

Considerations like those just discussed make it easy to understand why economists are biased in favor of growth. However, we are now entering a period where biological and physical constraints will soon put an end to economic growth.

Instead of burning our tropical forests, it might be wise for us to burn our books on growth-oriented economics! An entirely new form of economics is needed today - not the empty-world economics of Adam Smith, but what might be called "full-world economics", or "steady-state economics".

The present use of resources by the industrialized countries is extremely wasteful. A growing national economy must, at some point, exceed the real needs of the citizens. It has been the habit of the developed countries to create artificial needs by means of advertising, in order to allow economies to grow beyond the point where all real needs have been met; but this extra growth is wasteful, and in the future it will be important not to waste the earth's diminishing supply of non-renewable resources.

Thus, the times in which we live present a challenge: We need a revolution in economic thought. We must develop a new form of economics, taking into account the realities of the world's present situation - an economics based on real needs and on a sustainable equilibrium with the environment, not on the thoughtless assumption that growth can continue forever.

Adam Smith was perfectly correct in saying that the free market is the dynamo of economic growth; but exponential growth of human population and economic activity have brought us, in a surprisingly short time, from the empty-world situation in which he lived to a full-world situation. In today's world, we are pressing against the absolute limits of the earth's carrying capacity, and further growth carries with it the danger of future collapse. Full-world economics, the economics of the future, will no longer be able to rely on growth to give profits to stockbrokers or to solve problems of unemployment or to alleviate poverty. In the long run, growth of any kind is not sustainable; and we are now nearing the limits.

Like a speeding bus headed for a brick wall, the earth's rapidly-growing population of humans and its rapidly-growing economic activity are headed for a collision with a very solid barrier - the carrying capacity of the global environment. As in the case of the bus and the wall, the correct response to the situation is to apply the brakes in time - but fear prevents us from doing this. What will happen if we slow down very suddenly? Will not many of the passengers be injured? Undoubtedly. But what will happen if we hit the wall at full speed? Perhaps it would be wise, after all, to apply the brakes!

The memory of the great depression of 1929 makes us fear the consequences of an economic slowdown, especially since unemployment is already a serious problem in many parts of the world. Although the history of the 1929 depression is frightening, it may nevertheless be useful to look at the measures which were used then to bring the global economy back to its feet. A similar level of governmental responsibility may help us to avoid some of the more painful consequences of the necessary transition from the economics of growth to steady-state economics.

In the United States, President Franklin D. Roosevelt was faced with the difficult problems of the depression during his first few years in office. Roosevelt introduced a number of special governmental programs, such as the WPA, the Civilian Construction Corps and the Tennessee Valley Authority, which were designed to create new jobs on projects directed towards socially useful goals - building highways, airfields, auditoriums, harbors, housing projects, schools and dams. The English economist John Maynard Keynes, (1883-1946), provided an analysis of the factors that had caused the 1929 depression, and a theoretical justification of Roosevelt's policies.

The transition to a sustainable global society will require a similar level of governmental responsibility, although the measures needed are not the same as those which Roosevelt used to end the great depression. Despite the burst of faith in the free market which has followed the end of the Cold War, it seems unlikely that market mechanisms alone will be sufficient to solve problems of unemployment in the long-range future, or to achieve conservation of land, natural resources and environment.

9.2 Keynesian economics

In December, 1933, Keynes wrote to Franklin D. Roosevelt: "Dear Mr. President, You have made yourself the Trustee for those in every country who seek to mend the evils of our condition by reasoned experiment within the framework of the existing social system. If you fail, rational change will be gravely prejudiced throughout the world, leaving orthodoxy and revolution to fight it out. But if you succeed, new and bolder methods will

Fig. 9.1 *Franklin D. Roosevelt (1882-1945) with his dog Fala and Ruthie Bie at Hilltop in 1941. Roosevelt served as President of the United States from 1933 to 1945, and was starting his 4th term when he died. Although crippled by polio, he managed to convey an image of dynamism and confidence.*

be tried everywhere, and we may date the first chapter of a new economic era from your accession to office..."

"...Thus as the prime mover in the first stage of the technique of recovery I lay overwhelming emphasis on the increase of national purchasing power resulting from governmental expenditure which is financed by Loans and not by taxing present incomes. Nothing else counts in comparison with this. In a boom inflation can be caused by allowing unlimited credit to support the excited enthusiasm of business speculators. But in a slump governmental Loan expenditure is the only sure means of securing quickly a rising output at rising prices. That is why war has always caused intense industrial activity. In the past orthodox finance has regarded war as the only legitimate excuse for creating employment by governmental expenditure. You, Mr. President, having cast off such fetters, are free to engage in the interests of peace and prosperity the technique which hitherto has only been allowed to serve the purposes of war and destruction."

John Maynard Keynes (1883-1946), the author of this letter to Roosevelt, was the son of the Cambridge University economist and logician, Neville Keynes. After graduating from Eton and studying economics at King's College, Cambridge, Keynes spent a few years as a civil servant in the India Office. In 1909, he returned to Cambridge as a Fellow of King's College. He became a member of the "Bloomsbury Group", a collection of intellectual friends that included Virginia and Leonard Woolf, E.M. Forster, Clive and Vanessa Bell, Duncan Grant, Lytton Strachy, Roger Fry, and Bertrand Russell. In 1911, Keynes became the editor of the *Economic Journal*, a position that he retained almost until the end of his life.

In 1918, Keynes married the Russian ballerina Lydia Lopokova. They met at a party given by the Sitwells. Lydia was struggling to learn English, and one of her more interesting remarks was, "I dislike being in the country in August because my legs get so bitten by barristers". To everyone's surprise, Lydia proved to be the perfect wife for Keynes, encouraging his wide range of cultural interests. He and Lydia did much to develop the Cambridge Arts Theatre. Lydia maintained her interest in the ballet, although she no longer danced professionally. Visitors to the couple's house occasionally heard formidable thumpings from an upper room, and they realized that Lydia was practicing.

During World War I, Keynes worked in the British Treasury, helping to find ways to finance the war. In 1919, he was sent to the peace conference at Versailles as a representative of the Treasury. Keynes recognized the disastrous economic consequences that would follow from the Treaty of Versailles, and returning to Cambridge, he wrote *The Economic Consequences of the Peace* (1919). "It is an extraordinary fact", Keynes wrote, "that the fundamental problems of a Europe starving and disintegrating before their eyes, was the one question in which it was impossible to arouse the interest of the [Council of] Four."

The book became a best seller and was very influential in shaping public opinion, both in England and in the United States. In his book, Keynes predicted that the reparations imposed against Germany at Versailles would cause economic ruin. He advocated instead a loan system to rebuild postwar Europe. The plan advocated by Keynes was similar to the Marshall Plan that followed World War II. Had it been put into effect in 1919, it might have prevented the Second World War.

In 1936, Keynes published his magnum opus, *General Theory of Em-*

Fig. 9.2 *John Maynard Keynes (right) with Harry Dexter White at the Bretton Woods Conference. Keynes was an extremely tall man - 6 feet and 6 inches tall, i.e. 198 cm. Heart problems caused his early death.*

ployment, Interest and Money. In this book, he provided a theoretical explanation for the fact that the great depression showed no tendency to right itself, as well as arguments for governmental interventions to counter business cycles and to produce full employment. Once again, Keynes had

Fig. 9.3 *Migrant Mother, a photograph by Dorthea Lange, shows a destitute pea picker in California in 1936, during the Great Depression.*

written a best-seller. His *General Theory* proved to be one of the most influential books on economics ever written.

Keynes rebelled against the ideas of the classical economists, who believed that if let entirely alone, the world economy would correct itself. The classical economists recommended that, to end the depression, labor unions should be made illegal, minimum wages and long-term wage contracts abolished, and government spending curtailed (to restore business confidence). Then, they maintained, wages would fall, businessmen would hire more workers, and full employment and production would be restored. One reason for the popularity of the *General Theory* was that everyone knew the recommendations of the classical economists were bad policies. Now Keynes showed why these bad policies were also bad economics.

Keynes pointed out that a fall in wages would produce a fall in purchasing power, and hence a fall in aggregate demand. Producers would then be less able to sell their products. Thus Keynes believed that falling wages would deepen the depression, rather than ending it.

Part of Keynes' skepticism towards classical economics had to do with his criticisms of the short-term version of Say's Law, on which classical economics was based. In Chapter 2, we mentioned that Jean-Baptiste Say (1767-1832) believed a general glut to be impossible, since wages for the production of goods could be used by society to buy back its aggregate production. "A glut", Say wrote, "can take place only when there are too many means of production applied to one kind of product, and not enough to another."

Say considered the influence of the money supply on this process to be negligible, and he believed that the problem could be analyzed from the standpoint of barter. Say believed that no one would keep money for long. Having obtained money in a transaction, he believed, people would immediately spend it again. Thus Say did not worry about the problem of excessive saving that bothered both Malthus and Hobson.

"It is not the abundance of money", Say wrote, "but the abundance of other products in general that facilitates sales... Money performs no more than the role of a conduit in this double exchange. When the exchanges have been completed, it will be found that one has paid for products with products."

"It is worthwhile to remark", Say continued, "that a product is no sooner created than it, from that instant, affords a market for other products to the full extent of its value. When the producer has put the finishing hand to his product, he is most anxious to sell it immediately, lest its value should diminish in his hands. Nor is he less anxious to dispose of the money he may get for it; for the value of money is also perishable. But the only way to get rid of money is in the purchase some product or other. Thus the mere circumstance of creation of one product immediately opens a vent for other products."

Keynes disagreed with these conclusions in several respects. First of all, he did not believe, like Say, that the money supply played a negligible role in determining economic activity. Secondly he did not agree that the producer who has received money for his goods is necessarily "anxious to dispose of the money". As a recession deepens, the value of money in terms of goods increases, and therefore it is rational to keep money, hoping to get more goods for it at a later time. Whether it is more rational to keep money or to spend it immediately depends on the phase of the business cycle, Keynes pointed out.

In James Mill's version, Say's Law states that "supply creates its own demand". Keynes reversed this, and maintained in a depression, the fault may be on the demand side, i.e., "demand creates supply", rather than the reverse. It is true that during the great depression, many people were in need; but need does not constitute demand in the economic sense unless it is combined with purchasing power.

Keynes (like Malthus and Hobson) believed that excessive saving could be a serious problem, capable of causing a "general glut" or depression. By excessive saving, he meant saving beyond planned investment, a condition that could be caused by falling consumer demand, overinvestment in previous years, or lack of business confidence. The classical economists believed that excessive saving would be corrected by falling interest rates. Keynes did not believe that interest rates would respond quickly enough to perform this corrective function. Instead, Keynes believed, excessive savings would be in the end corrected by the fall in aggregate income which characterizes a recession or depression. The economy would reach a new equilibrium at low levels of employment, income, investment and production. This new, undesirable equilibrium would not be self-correcting. (By calling his theory a *General Theory*, Keynes meant that he treated not only the full-employment equilibrium, but also other types of equilibria.)

Keynes believed that active government fiscal and monetary policy could be effective in combating cycles of inflation and depression. *Fiscal policy* is defined as policy regarding government expenditure, while *monetary policy* means governmental policy with respect to the money supply. Keynes advocated a counter-cyclical use of these two tools, i.e. he believed that government spending and expansionist monetary policy should be used to combat recessions and depressions, while the opposite policies should be used to cool an economy whenever it became overheated.

Keynes visited Roosevelt in Washington in 1934. Roosevelt liked him, but found his theories overly mathematical. Nevertheless Keynes ideas influenced Roosevelt's policies, especially in 1937, when a new dip in the economy occurred. Over the years, Keynes' advocacy of counter-cyclical governmental intervention has become widely accepted, especially by social-democratic governments in Europe.

The New Deal measures inaugurated by Roosevelt were only partially effective in producing full employment. The reason that they were only partially successful was that although they were designed to help business get restarted, they were viewed with hostility by the business community. This hostility prevented Roosevelt from using fiscal policy on a large enough scale to produce full employment. Also, because businessmen felt uneasy with the new political climate, business investment remained sluggish.

One of the conclusions of Keynes' *General Theory* was that investment by expanding businesses is essential to keep an economy from contracting. This conclusion is worrying, because in the future, exponential expansion of business activity will gradually become less and less possible. Thus we can visualize a future need for governmental intervention to prevent a depression.

During World War II, Keynes advice on how to finance the war effort was sought by the British government. He did as much as he could, but his activity was limited by increasing heart problems. At the end of the war, Keynes represented England at the Breton Woods Conference, which established the World Bank and the International Monetary Fund. He received many honors - for example, he became Lord Keynes. However, his health remained unstable, and in 1946 he died of a heart attack. His life and work had produced a permanent change from the *laissez faire* economics of Adam Smith to an era of recognized governmental responsibility.

9.3 The transition to a sustainable economy

The Worldwatch Institute, Washington D.C., lists the following steps as necessary for the transition to sustainability[1]:

(1) Stabilizing population
(2) Shifting to renewable energy
(3) Increasing energy efficiency
(4) Recycling resources
(5) Reforestation
(6) Soil Conservation

All of these steps are labor-intensive; and thus, wholehearted governmental commitment to the transition to sustainability can help to solve the problem of unemployment.

In much the same spirit that Roosevelt (with Keynes' approval) used governmental powers to end the great depression, we must now urge our governments to use their powers to promote sustainability and to reduce the trauma of the transition to a steady-state economy. For example, an increase in the taxes on fossil fuels could make a number of renewable energy technologies economically competitive; and higher taxes on motor

[1]L.R. Brown and P. Shaw, 1982.

Fig. 9.4 *A reforestation project in Burkina Faso. Projects such as this may help the world to achieve sustainability, while simultaneously helping to solve problems of unemployment.*

fuels would be especially useful in promoting the necessary transition from private automobiles to bicycles and public transportation. Tax changes could also be helpful in motivating smaller families.

Governments already recognize their responsibility for education. In the future, they must also recognize their responsibility for helping young people to make a smooth transition from education to secure jobs. If jobs are scarce, work must be shared, in a spirit of solidarity, among those seeking employment; hours of work (and if necessary, living standards) must be reduced to insure a fair distribution of jobs. Market forces alone cannot achieve this. The powers of government are needed.

Economic activity is usually divided into two categories, 1) production of goods and 2) provision of services. It is the rate of production of goods that will be limited by the carrying capacity of the global environment. Services that have no environmental impact will not be constrained in this way. Thus a smooth transition to a sustainable economy will involve a shift of a large fraction the work force from the production of goods to the provision of services.

In his recent popular book *The Rise of the Creative Class*, the economist Richard Florida points out that in a number of prosperous cities - for example Stockholm - a large fraction of the population is already engaged in what might be called creative work - a type of work that uses few resources, and produces few waste products - work which develops knowledge and culture rather than producing material goods. For example, producing computer software requires few resources and results in few waste products. Thus it is an activity with a very small ecological footprint. Similarly, education, research, music, literature and art are all activities that do not weigh heavily on the carrying capacity of the global environment. Florida sees this as a pattern for the future, and maintains that everyone is capable of creativity. He visualizes the transition to a sustainable future economy as one in which a large fraction of the work force moves from industrial jobs to information-related work. Meanwhile, as Florida acknowledges, industrial workers feel uneasy and threatened by such trends.

9.4 Population and goods per capita

In the distant future, the finite carrying capacity of the global environment will impose limits on the amount of resource-using and waste-generating economic activity that it will be possible for the world to sustain. The consumption of goods per capita will be equal to this limited total economic activity divided by the number of people alive at that time. Thus, our descendants will have to choose whether they want to be very numerous and very poor, or less numerous and more comfortable, or very few and very rich. Perhaps the middle way will prove to be the best.

Given the fact that environmental carrying capacity will limit the sustainable level of resource-using economic activity to a fixed amount, average wealth in the distant future will be approximately inversely proportional to population over a certain range of population values.[2]

Suggestions for further reading

(1) R.L. Heilbroner, *The Worldly Philosophers, 5th edition*, Simon and Schuster, (1980).
(2) R. Harrod, *Life of John Maynard Keynes*, Harcourt, Brace, New York, (1951).

[2]Obviously, if the number of people is reduced to such an extent that it approaches zero, the average wealth will not approach infinity, since a certain level of population is needed to maintain a modern economy. However, if the global population becomes extremely large, the average wealth will indeed approach zero.

(3) J.M. Keynes, *Economic Consequences of the Peace*, Harcourt, Brace, New York, (1920).
(4) J.M. Keynes, *Essays in Persuasion*, Harcourt, Brace, New York, (1951).
(5) J.M. Keynes, *The General Theory of Employment, Interest and Money*, Harcourt, Brace, New York, (1964).
(6) R. Lekachman, *The Age of Keynes*, Random House, New York, (1966).
(7) R. Florida, *The Rise of the Creative Class*, Basic Books, (2002).
(8) H.E. Daly, *Steady-State Economics: The Economics of Biophysical Equilibrium and Moral Growth*, W.H. Freeman, San Francisco, (1977).
(9) H.E. Daly, *Steady-State Economics*, Island Press, Washington D.C., (1991).
(10) H.E. Daly, *Economics, Ecology and Ethics: Essays Towards a Steady-State Economy*, W.H. Freeman, San Francisco, (1980).
(11) H.E. Daly, *For the Common Good*, Beacon Press, Boston, (1989).
(12) Aspen Institute for Humanistic Studies, Program in International Affairs, *The Planetary Bargain*, Aspen, Colorado, (1975).
(13) W. Berry, *Home Economics*, North Point Press, San Francisco, (1987).
(14) L.R. Brown, *Building a Sustainable Society*, W.W. Norton, (1981).
(15) L.R. Brown, and P. Shaw, *Six Steps to a Sustainable Society*, Worldwatch Paper 48, Worldwatch Institute, Washington D.C., (1982).
(16) E. Eckholm, *Planting for the Future: Forestry for Human Needs*, Worldwatch Paper 26, Worldwatch Institute, Washington D.C., (1979).
(17) R. Goodland, H. Daly, S. El Serafy and B. von Droste (editors), *Environmentally Sustainable Economic Development: Building on Brundtland*, UNESCO, Paris, (1991).
(18) F. Hirsch, *Social Limits to Growth*, Harvard University Press, Cambridge, (1976).
(19) W. Leontief, et al., *The Future of the World Economy*, Oxford University Press, (1977).
(20) M. Lipton, *Why Poor People Stay Poor*, Harvard University Press, (1977).
(21) J. McHale, and M.C. McHale, *Basic Human Needs: A Framework for Action*, Center for Integrative Studies, Huston, (1977).
(22) D.L. Meadows, *Alternatives to Growth*, Ballinger, Cambridge, (1977).
(23) D.H. Meadows, *The Global Citizen*, Island Press, Washington D.C., (1991).
(24) D.L. Meadows, and D.H. Meadows (editors), *Toward Global Equilibrium*, Wright-Allen Press, Cambridge, Mass., (1973).
(25) L.W. Milbrath, *Envisioning a Sustainable Society*, State University of New York Press, Albany, (1989).

(26) R.E. Miles, *Awakening from the American Dream: The Social and Political Limits to Growth*, Universe Books, New York, (1976).
(27) S. Postel, and L. Heise, *Reforesting the Earth* , Worldwatch Paper 83, Worldwatch Institute, Washington D.C., (1988).
(28) M. Sagoff, *The Economy of the Earth*, Cambridge University Press, (1988).
(29) E.F. Schumacher, *Small is Beautiful: Economics As If People Mattered*, Harper and Row, New York, (1973).
(30) World Bank, *World Development Report*, Oxford University Press, New York, (published annually).
(31) G.P. Zachary, *A 'Green Economist' Warns Growth May Be Overrated*, The Wall Street Journal, June 25, (1996).
(32) H.E. Daly, *Sustainable Growth - An Impossibility Theorem*, Development, **3**, 45-47, (1990).
(33) H.E. Daly and K.N. Townsend, (editors), *Valuing the Earth. Economics, Ecology, Ethics*, MIT Press, Cambridge, Massachusetts, (1993)

Chapter 10

OPTIMUM GLOBAL POPULATION

10.1 The Green Revolution

Fig. 10.1 *Norman Borlaug. His work on developing high-yield disease-resistant plant varieties won him a Nobel Peace Prize in 1970.*

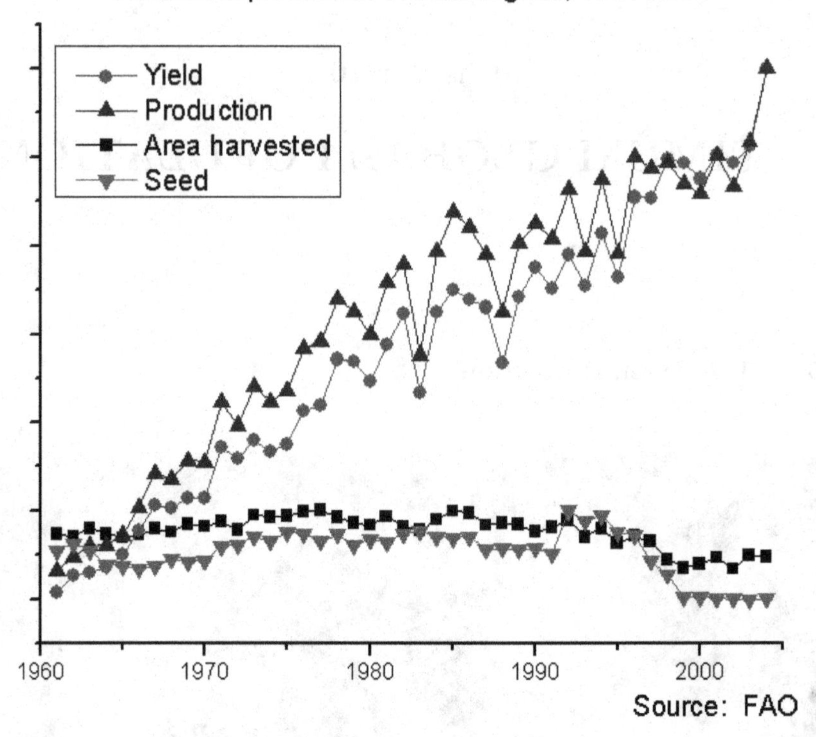

Fig. 10.2 *The total world production of coarse grain from 1961 to 2004. Although the yield greatly increased, the area under cultivation remained approximately constant. Between 1950 and 2005, wheat yields in Mexico increased from less than a ton per hectare-year to almost 5 tons, thanks to the work of Norman Borlaug, M.S. Swaminathan and other Green Revolution scientists.*

In 1944 the Norwegian-American plant geneticist Norman Borlaug was sent to Mexico by the Rockefeller Foundation to try to produce new wheat varieties that might increase Mexico's agricultural output. Borlaug's dedicated work on this project was spectacularly successful. He remained with the project for 16 years, and his group made 6,000 individual crossings of wheat varieties to produce high-yield disease-resistant strains.

In 1963, Borlaug visited India, bringing with him 100 kg. of seeds from each of his most promising wheat strains. After testing these strains in Asia, he imported 450 tons of the Lerma Rojo and Sonora 64 varieties - 250 tons for Pakistan and 200 for India. By 1968, the success of these varieties was so great that school buildings had to be commandeered to store the output. Borlaug's work began to be called a "Green Revolution". In India, the research on high-yield crops was continued and expanded by Prof. M.S. Swaminathan and his coworkers. The work of Green Revolution scientists, such Norman Borlaug and M.S. Swaminathan, has been credited with saving the lives of as many as a billion people.

Despite these successes, Borlaug believes that the problem of population growth is still a serious one. "Africa and the former Soviet republics", Borlaug states, "and the cerrado[1], are the last frontiers. After they are in use, the world will have no additional sizable blocks of arable land left to put into production, unless you are willing to level whole forests, which you should not do. So, future food-production increases will have to come from higher yields. And though I have no doubt that yields will keep going up, whether they can go up enough to feed the population monster is another matter. Unless progress with agricultural yields remains very strong, the next century will experience human misery that, on a sheer numerical scale, will exceed the worst of everything that has come before."

A very serious problem with Green Revolution plant varieties is that they require heavy inputs of pesticides, fertilizers and irrigation. Because of this, the use of high-yield varieties contributes to social inequality, since only rich farmers can afford the necessary inputs. Monocultures, such as the Green Revolution varieties may also prove to be vulnerable to future epidemics of plant diseases, such as the epidemic that caused the Irish Potato Famine in 1845. Even more importantly, pesticides, fertilizers and irrigation all depend on the use of fossil fuels. One must therefore ask whether high agricultural yields can be maintained in the future, when fossil fuels are expected to become prohibitively scarce and expensive.

10.2 Energy inputs of agriculture

Modern agriculture has become highly dependent on fossil fuels, especially on petroleum and natural gas. This is especially true of production of the high-yield grain varieties introduced in the Green Revolution, since these require especially large inputs of fertilizers, pesticides and irrigation. Today, fertilizers are produced using oil and natural gas, while pesticides are synthesized from petroleum feedstocks, and irrigation is driven by fossil fuel energy. Thus agriculture in the developed countries has become a process where inputs of fossil fuel energy are converted into food calories. If one focuses only on the farming operations, the fossil fuel energy inputs are distributed as follows:

[1] The cerrado is a large savanna region of Brazil.

(1) Manufacture of inorganic fertilizer, 31%
(2) Operation of field machinery, 19%
(3) Transportation, 16%
(4) Irrigation, 13%
(5) Raising livestock (not including livestock feed), 8%
(6) Crop drying, 5%
(7) Pesticide production, 5%
(8) Miscellaneous, 8%

The ratio of the fossil fuel energy inputs to the food calorie outputs depends on how many energy-using elements of food production are included in the accounting. David Pimentel and Mario Giampietro of Cornell University estimated in 1994 that U.S. agriculture required 0.7 kcal of fossil fuel energy inputs to produce 1.0 kcal of food energy. However, this figure was based on U.N. statistics that did not include fertilizer feedstocks, pesticide feedstocks, energy and machinery for drying crops, or electricity, construction and maintenance of farm buildings. A more accurate calculation, including these inputs, gives an input/output ratio of approximately 1.0. Finally, if the energy expended on transportation, packaging and retailing of food is included, Pimentel and Giampietro found the input/output ratio for the U.S. food system to be approximately 10, and this figure did not include energy used for cooking.

The Brundtland Report's [2] estimate of the global potential for food production assumes "that the area under food production can be around 1.5 billion hectares (3.7 billion acres - close to the present level), and that the average yields could go up to 5 tons of grain equivalent per hectare (as against the present average of 2 tons of grain equivalent)." In other words, the Brundtland Report assumes an increase in yields by a factor of 2.5. This would perhaps be possible if traditional agriculture could everywhere be replaced by energy-intensive modern agriculture using Green Revolution plant varieties. However, Pimentel and Giampietro's studies show that modern energy-intensive agricultural techniques cannot be maintained after fossil fuels have been exhausted.

At the time when the Brundtland Report was written (1987), the global average of 2 tons of grain equivalent per hectare included much higher yields from the sector using modern agricultural methods. Since energy-intensive petroleum-based agriculture cannot be continued in the post-fossil-fuel era, future average crop yields will probably be much less than 2 tons of grain equivalent per hectare.

---

[2] World Commission on Environment and Development, *Our Common Future*, Oxford University Press, (1987). This book is often called "The Brundtland Report" after Gro Harlem Brundtland, the head of WCED, who was then Prime Minister of Norway.

The 1987 global population was approximately 5 billion. This population was supported by 3 billion tons of grain equivalent per year. After fossil fuels have been exhausted, the total world agricultural output is likely to be considerably less than that, and therefore the population that it will be possible to support will probably be considerably less than 5 billion, assuming that our average daily per capita use of food calories remains the same, and assuming that the amount of cropland and pasturage remains the same (1.5 billion hectares cropland, 3.0 billion hectares pasturage).

The Brundtland Report points out that "The present (1987) global average consumption of plant energy for food, seed and animal feed amounts to 6,000 calories daily, with a range among countries of 3,000-15,000 calories, depending on the level of meat consumption." Thus there is a certain flexibility in the global population that can survive on a given total agricultural output. If the rich countries were willing to eat less meat, more people could be supported.

10.3 Limitations on cropland

With regard to the prospect of increasing the area of cropland, a report by the United Nations Food and Agricultural Organization (*Provisional Indicative World Plan for Agricultural Development*, FAO, Rome, 1970) states that "In Southern Asia,... in some countries of Eastern Asia, in the Near East and North Africa... there is almost no scope for expanding agricultural area... In the drier regions, it will even be necessary to return to permanent pasture the land that is marginal and submarginal for cultivation. In most of Latin America and Africa south of the Sahara, there are still considerable possibilities for expanding cultivated areas; but the costs of development are high, and it will often be more economical to intensify the utilization of areas already settled." Thus there is a possibility of increasing the area of cropland in Africa south of the Sahara and in Latin America, but only at the cost of heavy investment and at the additional cost of destruction of tropical rain forests.

Rather than an increase in the global area of cropland, we may encounter a future loss of cropland through soil erosion, salination, desertification, loss of topsoil, depletion of minerals in topsoil, urbanization and failure of water supplies. In China and in the southwestern part of the United States, water tables are falling at an alarming rate. The Ogallala aquifer (which supplies water to many of the plains states in the central and southern parts of the United States) has a large overdraft.

In the 1950's, both the U.S.S.R and Turkey attempted to convert arid grasslands into wheat farms. In both cases, the attempts were defeated by drought and wind erosion, just as the wheat farms of Oklahoma were overcome by drought and dust in the 1930's.

If irrigation of arid lands is not performed with care, salt may be deposited, so that the land is ruined for agriculture. This type of desertification can be seen, for example, in some parts of Pakistan. Another type of desertification can be seen in the Sahel region of Africa, south of the Sahara. Rapid population growth in the Sahel has led to overgrazing, destruction of trees, and wind erosion, so that the land has become unable to support even its original population.

Especially worrying is a prediction of the International Panel on Climate Change concerning the effect of global warming on the availability of water: According to Model A1 of the IPCC, global warming may, by the 2050's, have reduced by as much as 30% the water available in large areas of world that now a large producers of grain[3].

The earth's tropical rain forests are rapidly being destroyed for the sake of new agricultural land. Tropical rain forests are thought to be the habitat of more than half of the world's species of plants, animals and insects; and their destruction is accompanied by an alarming rate of extinction of species. The Harvard biologist, E.O. Wilson, estimates that the rate of extinction resulting from deforestation in the tropics may now exceed 4,000 species per year - 10,000 times the natural background rate (*Scientific American*, September, 1989).

The enormous biological diversity of tropical rain forests has resulted from their stability. Unlike northern forests, which have been affected by glacial epochs, tropical forests have existed undisturbed for millions of years. As a result, complex and fragile ecological systems have had a chance to develop. Professor Wilson expresses this in the following words:

"Fragile superstructures of species build up when the environment remains stable enough to support their evolution during long periods of time. Biologists now know that biotas, like houses of cards, can be brought tumbling down by relatively small perturbations in the physical environment. They are not robust at all."

The number of species which we have until now domesticated or used in medicine is very small compared with the number of potentially useful species still waiting in the world's tropical rain forests. When we destroy them, we damage our future. We ought to regard the annual loss of thousands of species as a tragedy, not only because biological diversity is potential wealth for human society, but also because every form of life deserves our respect and protection.

Every year, more than 100,000 square kilometers of rain forest are

[3]See the discussion of the Stern Report in Chapter 7.

cleared and burned, an area which corresponds to that of Switzerland and the Netherlands combined. Almost half of the world's tropical forests have already been destroyed. Ironically, the land thus cleared often becomes unsuitable for agriculture within a few years.

Tropical soils may seem to be fertile when covered with luxuriant vegetation, but they are usually very poor in nutrients because of leeching by heavy rains. The nutrients which remain are contained in the vegetation itself; and when the forest cover is cut and burned, the nutrients are rapidly lost.

Added to the agricultural and environmental problems, are problems of finance and distribution. Famines can occur even when grain is available somewhere in the world, because those who are threatened with starvation may not be able to pay for the grain, or for its transportation. The economic laws of supply and demand are not able to solve this type of problem. One says that there is no "demand" for the food (meaning demand in the economic sense), even though people are in fact starving.

Optimum population in the long-term future

What is the optimum population of the world? It is certainly not the maximum number that can be squeezed onto the globe by eradicating every species of plant and animal that cannot be eaten. The optimum global population is one that can be supported in comfort, equality and dignity - and with respect for the environment.

In 1848 (when there were just over one billion people in the world), John Stuart Mill described the optimal global population in the following words:

"The density of population necessary to enable mankind to obtain, in the greatest degree, all the advantages of cooperation and social intercourse, has, in the most populous countries, been attained. A population may be too crowded, although all be amply supplied with food and raiment."

"... Nor is there much satisfaction in contemplating the world with nothing left to the spontaneous activity of nature; with every rood of land brought into cultivation, which is capable of growing food for human beings; every flowery waste or natural pasture plowed up, all quadrupeds or birds which are not domesticated for man's use exterminated as his rivals for food, every hedgerow or superfluous tree rooted out, and scarcely a place left where a wild shrub or flower could grow without being eradicated as a weed in the name of improved agriculture. If the earth must lose that great portion of its pleasantness which it owes to things that the unlimited increase of wealth and population would extirpate from it, for the mere purpose of enabling it to support a larger, but not better or happier population, I sincerely hope, for the sake of posterity, that they will be content to be stationary, long before necessity compels them to it."[4]

Dennis Meadows, one of the authors of *Limits to Growth*, stated recently (in a private conversation) that the optimum human population in the distant future may be about 2 billion people.

But what about the near future? Will the global population of humans crash catastrophically after having exceeded the carrying capacity of the environment? There is certainly a danger that this will happen - a danger that the 21st century will bring very large scale famines to vulnerable parts of the world, because modern energy-intensive agriculture will be dealt a severe blow by prohibitively high petroleum prices. Climate change may also reduce the world's agricultural output. At present, there are only a few major food-exporting countries, notably the United States, Canada, Australia and Argentina. There is a danger that within a few decades, the United States will no longer be able to export food because of falling production and because of the demands of a growing population. We should be aware of these serious future problems if we are to have a chance of avoiding them.

10.4 The demographic transition

The developed industrial nations of the modern world have gone through a process known as the "demographic transition" - a shift from an equilibrium where population growth is held in check by the grim Malthusian forces of disease, starvation and war, to one where it is held in check by birth control and late marriage.

The transition begins with a fall in the death rate, caused by various factors, among which the most important is the application of scientific knowledge to the prevention of disease. Malthus gives the following list of some of the causes of high death rates: "...unwholesome occupations, severe labour and exposure to the seasons, extreme poverty, bad nursing of children, great towns, excesses of all kinds, the whole train of common diseases and epidemics, wars, plague and famine." The demographic transition begins when some of the causes of high death rates are removed.

Cultural patterns require some time to adjust to the lowered death rate, and so the birth rate continues to be high. Families continue to have six or seven children, just as they did when most of the children died before having children of their own. Therefore, at the start of the demographic transition, the population increases sharply. After a certain amount of time, however, cultural patterns usually adjust to the lowered death rate, and a

[4]John Stuart Mill, *Principles of Political Economy, With Some of Their Applications to Social Philosophy*, (1848).

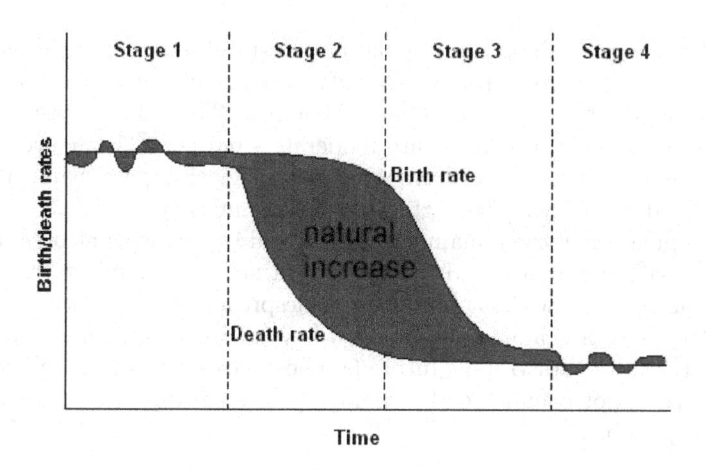

Fig. 10.3 *The demographic transition.*

new equilibrium is established, where both the birth rate and the death rate are low.

In Europe, this period of adjustment required about two hundred years. In 1750, the death rate began to fall sharply: By 1800, it had been cut in half, from 35 deaths per thousand people in 1750 to 18 in 1800; and it continued to fall. Meanwhile, the birth rate did not fall, but even increased to 40 births per thousand per year in 1800. Thus the number of children born every year was more than twice the number needed to compensate for the deaths!

By 1800, the population was increasing by more than two percent every year. In 1750, the population of Europe was 150 million; by 1800, it was roughly 220 million; by 1950 it had exceeded 540 million, and in 1970 it was 646 million.

Meanwhile the achievements of medical science and the reduction of the effects of famine and warfare had been affecting the rest of the world: In 1750, the non-European population of the world was only 585 million. By 1850 it had reached 877 million. During the century between 1850 and 1950, the population of Asia, Africa and Latin America more than doubled, reaching 1.8 billion in 1950. In the twenty years between 1950 and 1970, the population of Asia, Africa and Latin America increased still more sharply, and in 1970, this segment of the world's population reached 2.6 billion, bringing the world total to 3.6 billion. The fastest increase was in Latin America, where population almost doubled during the twenty years between 1950 and 1970.

The latest figures show that population has stabilized or in some cases is even decreasing in Europe, Russia, Canada, Japan, Cuba and New Zealand. In Argentina, the United States, China, Myanmar, Thailand and Australia, the rates of population increase are moderate - 0.6%-1.0%; but even this moderate rate of increase will have a heavy ecological impact, particularly in the United States, with its high rates of consumption.

The population of the remainder of the world is increasing at breakneck speed - 2%-4% per year - and it cannot continue to expand at this rate for very much longer without producing widespread famines, since modern intensive agriculture cannot be sustained beyond the end of the fossil fuel era. The threat of catastrophic future famines makes it vital that all countries that have not completed the demographic transition should do so as rapidly as possible.

10.5 Urbanization

The global rate of population growth has slowed from 2.0 percent per year in 1972 to 1.7 percent per year in 1987; and one can hope that it will continue to fall. However, it is still very high in most developing countries. For example, in Kenya, the population growth rate is 3.3 percent per year, which means that the population of Kenya will double in twenty-two years.

During the 60 years between 1920 and 1980 the urban population of the developing countries increased by a factor of 10, from 100 million to almost a billion. In 1950, the population of Sao Paulo in Brazil was 2.7 million. By 1980, it had grown to 12.6 million; and, if adjacent metropolitan areas such as Baixada Santista, Campinas, Sorocaba, etc. are included, Sao Paulo's present (2006) population is 29 million. Mexico City too has grown explosively to an unmanageable size. In 1950, the population of Mexico City was 3.05 million; in 1982 it was 16.0 million; and the population in 2000 was 17.8 million.

A similar explosive growth of cities can be seen in Africa and in Asia. In 1968, Lusaka, the capital of Zambia, and Lagos, the capital of Nigeria, were both growing at the rate of 14 percent per year, doubling in size every 5 years. In 1950, Nairobi, the capital of Kenya, had a population of 0.14 million. In a 1999 census, it was estimated to be between 3 and 4 million, having increased by a factor of 25.

In 1972, the population of Calcutta was 7.5 million. By the turn of

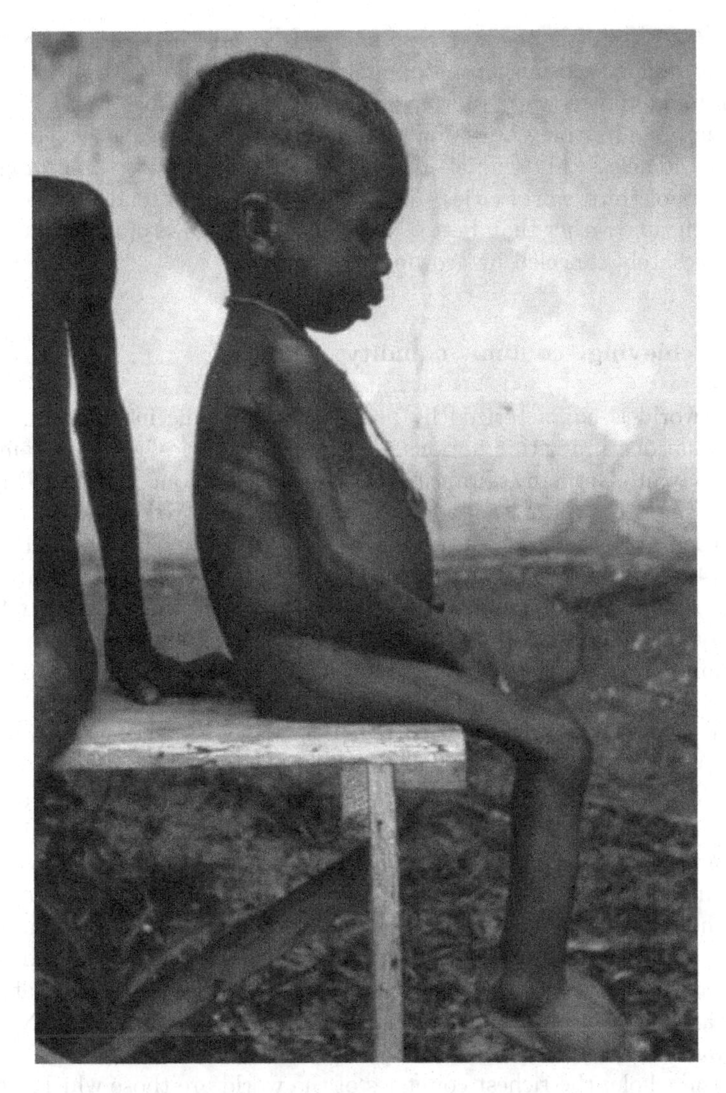

Fig. 10.4 *Because of the threat of widespread famine, it is vital that all countries should complete the demographic transition as quickly as possible.*

the century in 2000, it had almost doubled in size. This rapid growth produced an increase in the poverty and pollution from which Calcutta already suffered in the 1970's. The Hooghly estuary near Calcutta is choked with untreated industrial waste and sewage, and a large percentage of Calcutta's citizens suffer from respiratory diseases related to air pollution.

Governments in the third world, struggling to provide clean water, sanitation, roads, schools, medical help and jobs for all their citizens, are defeated by rapidly growing urban populations. Often the makeshift shantytowns inhabited by new arrivals have no piped water; or when water systems exist, the pressures may be so low that sewage seeps into the system.

Many homeless children, left to fend for themselves, sleep and forage in the streets of third world cities. These conditions have tended to become worse with time rather than better. Whatever gains governments can make are immediately canceled by growing populations.

10.6 Achieving economic equality

Today's world is characterized by intolerable economic inequalities, both between nations and within nations. A group of countries including (among others) Japan, Germany, France, the United Kingdom and the United States, has only 13% of the world's population, but receives 45% of the global PPP[5] income. By contrast, a second group, including 2.1 Billion people (45% of the world's population) receives only 9% of the global PPP income. Another indicator of inequality is the fact that the 50 million richest people in the world receive as much as the 2,700 million poorest.

18 million of our fellow humans die each year from poverty-related causes. Each year, 11 million children die before reaching their fifth birthday. 1.1 billion people live on less than 1$ per day; 2.7 billion live on less than 2$.

This degree of economic inequality is intolerable, not only for humanitarian reasons, but also because it contributes both directly and indirectly to the problem of war. A stable and peaceful world cannot be a world of glaring inequality. One indirect but important link between poverty and war is the following: The abolition of war will require effective global governance. However, because the enormous gap between the rich and poor of the world, rich nations block efforts to strengthen the United Nations. They fear that they would lose their privileged way of life if the UN were given (for example) the power to impose taxes.

On the whole, the richest countries of the world are those which industrialized early. During the 19th century and the early 20th century, these countries used their modern weapons to force other nations into the role of consumers of manufactured products and exporters of raw materials. The colonial era formally ended shortly after World War II, but colonialism was replaced by neocolonialism, and unsymmetrical economic relationships were perpetuated.

The way in which the industrialized countries maintain their control over

[5]Purchasing Power Parity

less developed nations can be illustrated by the "resource curse", i.e. the fact that resource-rich developing countries are no better off economically than those that lack resources, but are cursed with corrupt and undemocratic governments. This is because foreign corporations extracting local resources exist in a symbiotic relationship with corrupt local officials.

Another factor that perpetuates economic inequalities is foreign debt. At some time in the past, a typical developing country may have been induced to borrow money from banks in the industrialized part of the world. Often this money was used to buy armaments, whose economic effects were certainly not positive. The regime may now have changed, the weapons may be obsolete, but the debts remain. Payment of interest on foreign debts takes so much of the money of the developing countries that it overbalances the aid that they receive.

One might think that taxation of foreign resource-extracting firms would provide developing countries with large incomes. However, there is at present no international law governing multinational tax arrangements. These are usually agreed to on a bilateral basis, and the industrialized countries have stronger bargaining powers in arranging the bilateral agreements.

Endemic disease is strongly linked to poverty. Great improvements in reducing the effects of diseases like HIV/AIDS, malaria, shistosomiasis, trichoniasis, and river blindness could be made if pharmaceutical companies could be induced to do more research on tropical diseases and to provide drugs to developing countries at affordable prices. Other important measures would be universal vaccination programs, and the provision of safe water to all. It is in the interests of developed countries to promote health in the developing world, because air travel can quickly spread epidemics from one region to another.

Another important poverty-generating factor in the developing countries is war - often civil war. The five permanent members of the U.N. Security Council are, ironically, the five largest exporters of small arms. Small arms have a long life. The weapons poured into Africa by both sides during the Cold War are still there, and they contribute to political chaos and civil wars that block development and cause enormous human suffering.

The phrase "developing countries" is more than a euphemism; it expresses the hope that with the help of a transfer of technology from the industrialized nations, all parts of the world can achieve prosperity. Some of the forces that block this hope have just been mentioned. Another factor that prevents the achievement of worldwide prosperity is population growth.

In the words of Dr. Halfdan Mahler, former Director General of the World Health Organization, "Country after country has seen painfully achieved increases in total output, food production, health and educational facilities and employment opportunities reduced or nullified by excessive population growth."

The growth of population is linked to excessive urbanization, infrastructure failures and unemployment. In rural districts in the developing countries, family farms are often divided among a growing number of heirs until they can no longer be subdivided. Those family members who are no longer needed on the land have no alternative except migration to overcrowded cities, where the infrastructure is unable to cope so many new arrivals. Often the new migrants are forced to live in excrement-filled makeshift slums, where dysentery, hepatitis and typhoid are endemic, and where the conditions for human life sink to the lowest imaginable level. In Brazil, such shanty towns are called "favelas".

If modern farming methods are introduced in rural areas while population growth continues, the exodus to cities is aggravated, since modern techniques are less labor-intensive and favor large farms. In cities, the development of adequate infrastructure requires time, and it becomes a hopeless task if populations are growing rapidly. Thus, population stabilization is a necessary first step for development.

It can be observed that birth rates fall as countries develop. However, development is sometimes blocked by the same high birth rates that economic progress might have prevented. In this situation (known as the "demographic trap"), economic gains disappear immediately because of the demands of an exploding population.

For countries caught in the demographic trap, government birth control programs are especially important, because one cannot rely on improved social conditions to slow birth rates. Since health and lowered birth rates should be linked, it is appropriate that family-planning should be an important part of programs for public health and economic development.

A recent study conducted by Robert F. Lapham of Demographic Health Surveys and W. Parker Mauldin of the Rockefeller Foundation has shown that the use of birth control is correlated both with socio-economic setting and with the existence of strong family-planning programs. The implication of this study is that even in the absence of increased living standards, family-planning programs can be successful, provided they have strong government support. Among the developing countries with vigorous and successful family planning programs are Indonesia, Colombia and Thailand. In Colombia, fertility has fallen over the past 30 years from 7.1 children per woman to 2.9; while in Thailand, it fell from 6.5 children per woman to 2.1 The examples of Japan, Singapore and Hong Kong show that countries which have stabilized their populations have found it easy, thereafter, to raise living standards.

China, the world's most populous nation, has adopted the somewhat draconian policy of allowing only one child per family. This policy has, until now, been most effective in towns and cities; but with time it may also become effective in rural areas. Chinese leaders obtained popular support for their one-child policy by means of an educational program which emphasized future projections of diminishing water resources and diminishing cropland per person if population increased unchecked. Like other developing countries, China has a very young population, which will continue to grow even when fertility has fallen below the replacement level because so many of its members are contributing to the birth rate rather than to the death rate. China's present population is 1.3 billion. Its projected population for the year 2025 is 1.5 billion.

Education of women and higher status for women are vitally important measures, not only for their own sake, but also because in many countries these social reforms have proved to be the key to lower birth rates. Religious leaders who oppose programs for the education of women and for family planning on "ethical" grounds should think carefully about the scope and consequences of the catastrophic global famine which will undoubtedly occur within the next 50 years if population is allowed to increase unchecked.

At the United Nations Conference on Population and Development, held in Cairo in September, 1994, a theme which emerged very clearly was that one of the most important keys to controlling the global population explosion is giving women better education and equal rights. These goals are desirable for the sake of increased human happiness, and for the sake of the uniquely life-oriented point of view which women can give us; but in addition, education and improved status for women have shown themselves to be closely connected with lowered birth rates. When women lack education and independent careers outside the home, they can be forced into the role of baby-producing machines by men who do not share in the drudgery of cooking, washing and cleaning; but when women have educational, legal, economic, social and political equality with men, experience has shown that they choose to limit their families to a moderate size.

Today's industrialized countries were fortunate in undergoing their development during an era when fossil fuels were cheap and plentiful and when the global environment was not so stressed as it is today. In the future, development must always be qualified by the word "sustainable", and it must take place in an era of energy scarcity and resource scarcity. Thus economic progress in the developing nations must follow a different path than that which was followed historically by the present industrialized countries. The developing nations will be aided by recent progress in science and technology, and hopefully they will be able to avoid the worst

Fig. 10.5 *Education of women and higher status for women are vitally important measures, not only for their own sake, but also because these social reforms have proved to be the key to lower birth rates.*

features of the 19th century Industrial Revolution in Europe; but under all circumstances, a necessary precondition must be population stabilization.

Suggestions for further reading

(1) L.R. Brown, *Who Will Feed China?*, W.W. Norton, New York, (1995).
(2) H. Hanson, N.E. Borlaug and N.E. Anderson, *Wheat in the Third World*, Westview Press, Boulder, Colorado, (1982).

(3) A. Dil, ed., *Norman Borlaug and World Hunger*, Bookservice International, San Diego/Islamabad/Lahore, (1997).
(4) N.E. Borlaug, *The Green Revolution Revisited and the Road Ahead*, Norwegian Nobel Institute, Oslo, Norway, (2000).
(5) N.E. Borlaug, *Ending World Hunger. The Promise of Biotechnology and the Threat of Antiscience Zealotry*, Plant Physiology, **124**, 487-490, (2000).
(6) M. Giampietro and D. Pimentel, *The Tightening Conflict: Population, Energy Use and the Ecology of Agriculture*, in *Negative Population Forum*, L. Grant ed., Negative Population Growth, Inc., Teaneck, N.J., (1993).
(7) H.W. Kendall and D. Pimentel, *Constraints on the Expansion of the Global Food Supply*, Ambio, **23**, 198-2005, (1994).
(8) D. Pimentel et al., *Natural Resources and Optimum Human Population*, Population and Environment, **15**, 347-369, (1994).
(9) D. Pimental et al., *Environmental and Economic Costs of Soil Erosion and Conservation Benefits*, Science, **267**, 1117-1123, (1995).
(10) RS and NAS, *The Royal Society and the National Academy of Sciences on Population Growth and Sustainability*, Population and Development Review, **18**, 375-378, (1992).
(11) A.M. Altieri, *Agroecology: The Science of Sustainable Agriculture*, Westview Press, Boulder, Colorado, (1995).
(12) L.R. Brown, *Seeds of Change*, Praeger Publishers, New York, (1970).
(13) G. Conway, *The Doubly Green Revolution*, Cornell University Press, (1997).
(14) J. Dreze and A. Sen, *Hunger and Public Action*, Oxford University Press, (1991).
(15) T. Berry, *The Dream of the Earth*, Sierra Club Books, San Francisco, (1988).
(16) G. Bridger, and M. de Soissons, *Famine in Retreat?*, Dent, London, (1970).
(17) L.R. Brown, *The Twenty-Ninth Day*, W.W. Norton, New York, (1978).
(18) L.R. Brown, *The Worldwide Loss of Cropland*, Worldwatch Paper 24, Worldwatch Institute, Washington, D.C., (1978).
(19) L.R. Brown, and J.L. Jacobson, *Our Demographically Divided World*, Worldwatch Paper 74, Worldwatch Institute, Washington D.C., (1986).
(20) L.R. Brown, and J.L. Jacobson, *The Future of Urbanization: Facing the Ecological and Economic Constraints*, Worldwatch Paper 77, Worldwatch Institute, Washington D.C., (1987).

(21) L.R. Brown, and others, *State of the World*, W.W. Norton, New York, (published annually).
(22) W. Brandt, *World Armament and World Hunger: A Call for Action*, Victor Gollanz Ltd., London, (1982).
(23) W.V. Chandler, *Materials Recycling: The Virtue of Necessity*, Worldwatch Paper 56, Worldwatch Institute, Washington D.C, (1983).
(24) A.K.M.A. Chowdhury and L.C. Chen, *The Dynamics of Contemporary Famine*, Ford Foundation, Dacca, Pakistan, (1977).
(25) M.E. Clark, *Ariadne's Thread: The Search for New Modes of Thinking*, St. Martin's Press, New York, (1989).
(26) W.C. Clark and others, *Managing Planet Earth*, Special Issue, *Scientific American*, September, (1989).
(27) J.-C. Chesnais, *The Demographic Transition*, Oxford, (1992).
(28) C.M. Cipola, *The Economic History of World Population*, Penguin Books Ltd., (1974).
(29) B. Commoner, *The Closing Circle: Nature, Man and Technology*, Bantam Books, New York, (1972).
(30) Council on Environmental Quality and U.S. Department of State, *Global 2000 Report to the President: Entering the Twenty-First Century*, Technical Report, Volume 2, U.S. Government Printing Office, Washington D.C., (1980).
(31) A.B. Durning, *Action at the Grassroots: Fighting Poverty and Environmental Decline*, Worldwatch Paper , Worldwatch Institute, Washington D.C., (1989).
(32) P. Donaldson, *Worlds Apart: The Economic Gulf Between Nations*, Penguin Books Ltd., (1973).
(33) J.C.I. Dooge et al. (editors), *Agenda of Science for Environment and Development into the 21st Century*, Cambridge University Press, (1993).
(34) E. Draper, *Birth Control in the Modern World*, Penguin Books, Ltd., (1972).
(35) Draper Fund Report No. 15, *Towards Smaller Families: The Crucial Role of the Private Sector*, Population Crisis Committee, 1120 Nineteenth Street, N.W., Washington D.C. 20036, (1986).
(36) E. Eckholm, *Losing Ground: Environmental Stress and World Food Prospects*, W.W. Norton, New York, (1975).
(37) E. Eckholm, *The Picture of Health: Environmental Sources of Disease*, New York, (1976).
(38) Economic Commission for Europe, *Air Pollution Across Boundaries*, United Nations, New York, (1985).
(39) A.H. Ehrlich and U. Lele, *Humankind at the Crossroads: Building*

a Sustainable Food System, in *Draft Report of the Pugwash Study Group: The World at the Crossroads*, Berlin, (1992).
(40) P.R. Ehrlich, *The Population Bomb*, Sierra/Ballentine, New York, (1972).
(41) P.R. Ehrlich, A.H. Ehrlich and J. Holdren, *Human Ecology*, W.H. Freeman, San Francisco, (1972).
(42) P.R. Ehrlich, A.H. Ehrlich and J. Holdren, *Ecoscience: Population, Resources, Environment*, W.H. Freeman, San Francisco, (1977)
(43) P.R. Ehrlich and A.H. Ehrlich, *Extinction*, Victor Gollancz, London, (1982).
(44) P.R. Ehrlich and A.H. Ehrlich, *Healing the Planet*, Addison Wesley, Reading MA, (1991).
(45) P.R. Ehrlich and A.H. Ehrlich, *The Population Explosion*, Arrow Books, (1991).
(46) Food and Agricultural Organization, *The State of Food and Agriculture*, United Nations, Rome, (published annually).
(47) K. Griffin, *Land Concentration and Rural Poverty*, Holmes and Meyer, New York, (1976).
(48) G. Hagman and others, *Prevention is Better Than Cure*, Report on Human Environmental Disasters in the Third World, Swedish Red Cross, Stockholm, Stockholm, (1986).
(49) M. ul Haq, *The Poverty Curtain: Choices for the Third World*, Columbia University Pres, New York, (1976).
(50) G. Hardin, "The Tragedy of the Commons", *Science*, December 13, (1968).
(51) E. Havemann, *Birth Control*, Time-Life Books, (1967).
(52) J. Jacobsen, *Promoting Population Stabilization: Incentives for Small Families*, Worldwatch Paper 54, Worldwatch Institute, Washington D.C., (1983).
(53) N. Keyfitz, *Applied Mathematical Demography*, Wiley, New York, (1977).
(54) W. Latz (ed.), *Future Demographic Trends*, Academic Press, New York, (1979).
(55) H. Le Bras, *La Planète au Village*, Datar, Paris, (1993).
(56) E. Mayr, *Population, Species and Evolution*, Harvard University Press, Cambridge, (1970).
(57) N. Myers, *The Sinking Ark*, Pergamon, New York, (1972).
(58) N. Myers, *Conservation of Tropical Moist Forests*, National Academy of Sciences, Washington D.C., (1980).
(59) K. Newland, *Infant Mortality and the Health of Societies*, Worldwatch Paper 47, Worldwatch Institute, Washington D.C., (1981).

(60) W. Ophuls, *Ecology and the Politics of Scarcity*, W.H. Freeman, San Francisco, (1977).
(61) D.W. Orr, *Ecological Literacy*, State University of New York Press, Albany, (1992).
(62) A. Peccei, *The Human Quality*, Pergamon Press, Oxford, (1977).
(63) A. Peccei, *One Hundred Pages for the Future*, Pergamon Press, New York, (1977).
(64) A. Peccei and D. Ikeda, *Before it is Too Late*, Kodansha International, Tokyo, (1984).
(65) E. Pestel, *Beyond the Limits to Growth*, Universe Books, New York, (1989).
(66) D.C. Pirages and P.R. Ehrlich, *Ark II: Social Responses to Environmental Imperatives*, W.H. Freeman, San Francisco, (1974).
(67) Population Reference Bureau, *World Population Data Sheet*, PRM, 777 Fourteenth Street NW, Washington D.C. 20007, (published annually).
(68) R. Pressat, *Population*, Penguin Books Ltd., (1970).
(69) M. Rechcigl (ed.), *Man/Food Equation*, Academic Press, New York, (1975).
(70) J.C. Ryan, *Life Support: Conserving Biological Diversity*, Worldwatch Paper 108, Worldwatch Institute, Washington D.C., (1992).
(71) J. Shepard, *The Politics of Starvation*, Carnegie Endowment for International Peace, Washington D.C., (1975).
(72) P.B. Smith, J.D. Schilling and A.P. Haines, *Introduction and Summary*, in *Draft Report of the Pugwash Study Group: The World at the Crossroads*, Berlin, (1992).
(73) B. Stokes, *Local Responses to Global Problems: A Key to Meeting Basic Human Needs*, Worldwatch Paper 17, Worldwatch Institute, Washington D.C., (1978).
(74) L. Timberlake, *Only One Earth: Living for the Future*, BBC/ Earthscan, London, (1987).
(75) UNEP, *Environmental Data Report*, Blackwell, Oxford, (published annually).
(76) UNESCO, *International Coordinating Council of Man and the Biosphere*, MAB Report Series No. 58, Paris, (1985).
(77) United Nations Fund for Population Activities, *A Bibliography of United Nations Publications on Population*, United Nations, New York, (1977).
(78) United Nations Fund for Population Activities, *The State of World Population*, UNPF, 220 East 42nd Street, New York, 10017, (published annually).

(79) United Nations Secretariat, *World Population Prospects Beyond the Year 2000*, U.N., New York, (1973).
(80) J. van Klinken, *Het Dierde Punte*, Uitgiversmaatschappij J.H. Kok-Kampen, Netherlands (1989).
(81) P.M. Vitousek, P.R. Ehrlich, A.H. Ehrlich and P.A. Matson, *Human Appropriation of the Products of Photosynthesis*, Bioscience, *34*, 368-373, (1986).
(82) B. Ward and R. Dubos, *Only One Earth*, Penguin Books Ltd., (1973).
(83) WHO/UNFPA/UNICEF, *The Reproductive Health of Adolescents: A Strategy for Action*, World Health Organization, Geneva, (1989).
(84) E.O. Wilson, *Sociobiology*, Harvard University Press, (1975).
(85) E.O. Wilson (ed.), *Biodiversity*, National Academy Press, Washington D.C., (1988).
(86) E.O. Wilson, *The Diversity of Life*, Allen Lane, The Penguin Press, London, (1992).
(87) G. Woodwell (ed.), *The Earth in Transition: Patterns and Processes of Biotic Impoverishment*, Cambridge University Press, (1990).
(88) World Commission on Environment and Development, *Our Common Future*, Oxford University Press, (1987).
(89) World Bank, *Poverty and Hunger: Issues and Options for Food Security in Developing Countries*, Washington D.C., (1986).
(90) World Resources Institute (WRI), *Global Biodiversity Strategy*, The World Conservation Union (IUCN), United Nations Environment Programme (UNEP), (1992).
(91) World Resources Institute, *World Resources*, Oxford University Press, New York, (published annually).
(92) J.E. Young et al., *Mining the Earth* , Worldwatch Paper 109, Worldwatch Institute, Washington D.C., (1992).
(93) J.E. Cohen, *How Many People Can the Earth Support?*, W.W. Norton, New York, (1995).
(94) D.W. Pearce and R.K. Turner, *Economics of Natural Resources and the Environment*, Johns Hopkins University Press, Baltimore, (1990).
(95) P. Bartelmus, *Environment, Growth and Development: The Concepts and Strategies of Sustainability*, Routledge, New York, (1994).
(96) D. Pimental et al., *Natural Resources and Optimum Human Population*, Population and Environment, **15**, 347-369, (1994).
(97) D. Pimentel and M. Pimentel, *Food Energy and Society*, University Press of Colorado, Niwot, Colorado, (1996).
(98) H. Brown, *The Human Future Revisited. The World Predicament and Possible Solutions*, W.W. Norton, New York, (1978).

(99) W. Jackson, *Man and the Environment*, Wm. C. Brown, Dubuque, Iowa, (1971).

(100) L.R. Brown, et al., *Saving the Planet. How to Shape and Environmentally Sustainable Global Economy*, W.W. Norton, New York, (1991).

(101) L.R. Brown, *Postmodern Malthus: Are There Too Many of Us to Survive?*, The Washington Post, July 18, (1993).

(102) L.R. Brown and H. Kane, *Full House. Reassessing the Earth's Population Carrying Capacity*, W.W. Norton, New York, (1991).

(103) J. Amos, *Climate Food Crisis to Deepen*, BBC News (5 September, 2005).

(104) J. Vidal and T. Ratford, *One in Six Countries Facing Food Shortage*, The Guardian, (30 June, 2005).

(105) J. Mann, *Biting the Environment that Feeds Us*, The Washington Post, July 29, 1994.

(106) G.R. Lucas, Jr., and T.W. Ogletree, (editors), *Lifeboat Ethics. The Moral Dilemmas of World Hunger*, Harper and Row, New York.

(107) J.L. Jacobson, *Gender Bias: Roadblock to Sustainable Development*, Worldwatch Institute, Washington D.C., (1992).

Chapter 11

THE PROBLEM OF WAR

11.1 The passions of mankind

The explosion of human knowledge

Cultural evolution depends on the non-genetic storage, transmission, diffusion and utilization of information. The development of human speech, the invention of writing, the development of paper and printing, and finally in modern times, mass media, computers and the Internet - all these have been crucial steps in society's explosive accumulation of information and knowledge. Human cultural evolution proceeds at a constantly-accelerating speed, so great in fact that it threatens to shake society to pieces.

Every species changes gradually through genetic evolution; but with humans, cultural evolution has rushed ahead with such a speed that it has completely outstripped the slow rate of genetic change. Genetically we are almost identical with our neolithic ancestors, but their world has been replaced by a world of quantum theory, relativity, supercomputers, antibiotics, genetic engineering and space telescopes - unfortunately also a world of nuclear weapons and nerve gas.

Because of the slowness of genetic evolution in comparison to the rapid and constantly-accelerating rate of cultural change, our bodies and emotions (as Malthus put it, the "passions of mankind") are not adapted to our new way of life. They still reflect the way of life of our hunter-gatherer ancestors.

Within rapidly-moving cultural evolution, we can observe that technical change now moves with such astonishing rapidity that neither social institutions, nor political structures, nor education, nor public opinion can keep pace. The lightning-like pace of technical progress has made many of our ideas and institutions obsolete. For example, the absolutely-sovereign nation-state and the institution of war have both become dangerous anachronisms in an era of instantaneous communication, global interdependence and all-destroying weapons.

In many respects, human cultural evolution can be regarded as an enormous success. However, at the start of the 21st century, most thoughtful observers agree that civilization is entering a period of crisis. As all curves move exponentially upward - population, production, consumption, rates of scientific discovery, and so on - one can observe signs of increasing environmental stress, while the continued existence and spread of nuclear weapons threatens civilization with destruction. Thus while the explosive growth of knowledge has brought many benefits, the problem of achieving a stable, peaceful and sustainable world remains serious and challenging.

Tribal emotions and nationalism

In discussing conflicts, we must be very careful to distinguish between two distinct types of aggression exhibited by both humans and animals. The first is intra-group aggression, which is often seen in rank-determining struggles, for example when two wolves fight for pack leadership, or when males fight for the privilege of mating with females. Another, completely different, type of aggression is seen when a group is threatened by outsiders. Most animals, including humans, then exhibit a communal defense response - self-sacrificing and heroic combat against whatever is perceived to be an external threat. It is this second type of aggression that makes war possible.

Arthur Koestler has described inter-group aggression in an essay entitled *The Urge to Self-Destruction* [1], where he writes: "Even a cursory glance at history should convince one that individual crimes, committed for selfish motives, play a quite insignificant role in the human tragedy compared with the numbers massacred in unselfish love of one's tribe, nation, dynasty, church or ideology... Wars are not fought for personal gain, but out of loyalty and devotion to king, country or cause..."

"We have seen on the screen the radiant love of the Führer on the faces of the Hitler Youth... They are transfixed with love, like monks in ecstasy on religious paintings. The sound of the nation's anthem, the sight of its proud flag, makes you feel part of a wonderfully loving community. The fanatic is prepared to lay down his life for the object of his worship, as the lover is prepared to die for his idol. He is, alas, also prepared to kill anybody who represents a supposed threat to the idol." The emotion described here by Koestler is the same as the communal defense mechanism ("militant enthusiasm") described below in biological terms by the Nobel Laureate ethologist Konrad Lorenz.

In *On Aggression*, Lorenz gives the following description of the emotions

[1] In *The Place of Value in a World of Facts*, A. Tiselius and S. Nielsson editors, Wiley, New York, (1970).

of a hero preparing to risk his life for the sake of the group: "In reality, militant enthusiasm is a specialized form of communal aggression, clearly distinct from and yet functionally related to the more primitive forms of individual aggression. Every man of normally strong emotions knows, from his own experience, the subjective phenomena that go hand in hand with the response of militant enthusiasm. A shiver runs down the back and, as more exact observation shows, along the outside of both arms. One soars elated, above all the ties of everyday life, one is ready to abandon all for the call of what, in the moment of this specific emotion, seems to be a sacred duty. All obstacles in its path become unimportant; the instinctive inhibitions against hurting or killing one's fellows lose, unfortunately, much of their power. Rational considerations, criticisms, and all reasonable arguments against the behavior dictated by militant enthusiasm are silenced by an amazing reversal of all values, making them appear not only untenable, but base and dishonorable. Men may enjoy the feeling of absolute righteousness even while they commit atrocities. Conceptual thought and moral responsibility are at their lowest ebb. As the Ukrainian proverb says: 'When the banner is unfurled, all reason is in the trumpet'."

"The subjective experiences just described are correlated with the following objectively demonstrable phenomena. The tone of the striated musculature is raised, the carriage is stiffened, the arms are raised from the sides and slightly rotated inward, so that the elbows point outward. The head is proudly raised, the chin stuck out, and the facial muscles mime the 'hero face' familiar from the films. On the back and along the outer surface of the arms, the hair stands on end. This is the objectively observed aspect of the shiver!"

"Anybody who has ever seen the corresponding behavior of the male chimpanzee defending his band or family with self-sacrificing courage will doubt the purely spiritual character of human enthusiasm. The chimp, too, sticks out his chin, stiffens his body, and raises his elbows; his hair stands on end, producing a terrifying magnification of his body contours as seen from the front. The inward rotation of the arms obviously has the purpose of turning the longest-haired side outward to enhance the effect. The whole combination of body attitude and hair-raising constitutes a bluff. This is also seen when a cat humps its back, and is calculated to make the animal appear bigger and more dangerous than it really is. Our shiver, which in German poetry is called a 'heiliger Schauer', a 'holy' shiver, turns out to be the vestige of a prehuman vegetative response for making a fur bristle which we no longer have. To the humble seeker for biological truth, there cannot be the slightest doubt that human militant enthusiasm evolved out of a communal defense response of our prehuman ancestor."

Lorenz goes on to say, "An impartial visitor from another planet, looking at man as he is today - in his hand the atom bomb, the product of his intelligence - in his heart the aggression drive, inherited from his anthropoid ancestors, which the same intelligence cannot control - such a visitor would not give mankind much chance of survival."

Members of tribe-like groups are bound together by strong bonds of altruism and loyalty. Echos of these bonds can be seen in present-day family groups, in team sports, in the fellowship of religious congregations, and in the bonds that link soldiers to their army comrades and to their nation.

Warfare involves not only a high degree of aggression, but also an extremely high degree of altruism. Soldiers kill, but they also sacrifice their own lives. Thus patriotism and duty are as essential to war as the willingness to kill.

Tribalism involves passionate attachment to one's own group, self-sacrifice for the sake of the group, willingness both to die and to kill if necessary to defend the group from its enemies, and belief that in case of a conflict, one's own group is always in the right. Unfortunately these emotions make war possible; and today a Third World War might lead to the destruction of civilization.

Formation of group identity in modern nations

Our hunter-gatherer ancestors lived together in small genetically-homogeneous tribes, but modern nations are not only much larger, but also often multiracial and multiethnic.

If we think, for example of very large nations like the United States, Brazil, China or India, we are struck by the fact that it has been possible to achieve social cohesion within them, despite the size and ethnic diversity of these enormous countries.

If it is possible to produce social coherence over such large land areas and such with such multicultural populations, it must also be possible to make the entire world function as a single society, bound together by bonds of loyalty to humanity as a whole - a new kind of loyalty that transcends the old boundaries between nations.

Nevertheless, since a tendency towards tribalism seems to be part of human nature, we should not underestimate the difficulty of making an ethnically diverse society function as a single unit. We can and must achieve a united world, where human solidarity transcends all boundaries, but this will require commitment and responsibility on the part of religions, mass media, educational systems and governments throughout the world.

The same techniques that are today used to produce patriotism and solidarity within ethnically diverse nations must be used produce a higher loyalty to all of humanity. One can continue to feel loyalty to one's own family, ethnic group and nation, and at the same time recognize that the well-being of humanity as a whole represents a still higher goal. Loyalty to the entire human race is a higher form of patriotism.

Fig. 11.1 *"The Third of May, 1808: The Execution of the Defenders of Madrid", by Francesco Goya.*

11.2 Modern weapons

Hiroshima and Nagasaki

On August 6, 1945, a nuclear bomb was used against the predominantly civilian population of the city of Hiroshima in an already virtually defeated Japan. A few days later, a similar bomb was used against Nagasaki. The population of Hiroshima was 250,000, and of these, roughly 100,000 were killed immediately while another 100,000 were injured. Large numbers died later from radiation sickness, so that the total number of deaths reached approximately 140,000.

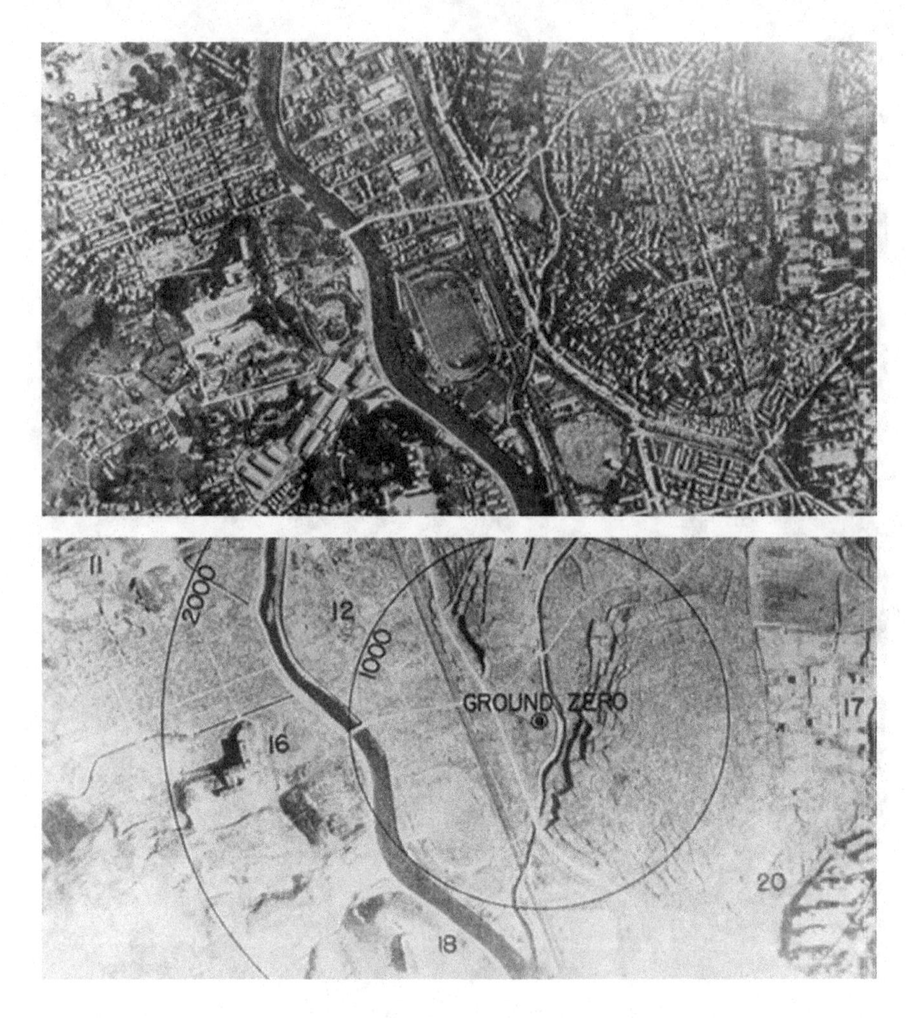

Fig. 11.2 *Nagasaki before and after the nuclear explosion.*

Many people were trapped in the wreckage of their houses and burned to death in the fires that followed the two nuclear explosions. In places near to the epicenters, people were entirely vaporized, leaving only shadows burned onto the pavement to show where they had been.

The postwar nuclear arms race

Although there was relief that the Second World War was over, many thoughtful people, including the French author Albert Camus, were horrified that bombs of this type had been used against civilian populations. Shortly after the bombings, Camus wrote: "Our technical civilization has just reached its greatest level of savagery. We will have to choose, in the more or less near future, between collective suicide and the intelligent use of our scientific conquests."

The USSR exploded its first nuclear bomb on August 29, 1949. The Soviet bomb had a yield equivalent to 21,000 tons of TNT, and it had been made from plutonium produced in a nuclear reactor. The United Kingdom also began to build its own nuclear weapons.

The explosion of the Soviet nuclear bomb caused feelings of panic in the United States, and President Truman authorized an all-out effort to build superbombs using thermonuclear reactions - the reactions that heat the sun and stars. On October 31, 1952, the first US thermonuclear device was exploded at Eniwetok Atoll in the Pacific Ocean. It had a yield of 10.4 megatons, that is to say it had an explosive power equivalent to ten million tons of TNT. Thus the first thermonuclear bomb was seven hundred times as powerful as the bombs that had devastated Hiroshima and Nagasaki. In March, 1954, the US tested another hydrogen bomb at the Bikini Atoll in the Pacific Ocean. It was more than 1000 times more powerful than the Hiroshima bomb. The Japanese fishing boat, Lucky Dragon, was 135 kilometers from the Bikini explosion, but radioactive fallout from the test killed one crew member and made all the others seriously ill.

In England, Prof. Joseph Rotblat, a Polish scientist who had resigned from the Manhattan Project for for moral reasons when it became clear that Germany would not develop nuclear weapons, was asked to appear on a BBC program to discuss the Bikini test. He was asked to discuss the technical aspects of H-bombs, while the Archbishop of Canterbury and the philosopher Bertrand Russell were invited to discuss the moral aspects.

Rotblat had became convinced that the Bikini bomb must have involved a third stage, where fast neutrons from the hydrogen thermonuclear reaction produced fission in a casing of ordinary uranium. Such a bomb would produce enormous amounts of highly dangerous radioactive fallout, and Rotblat became extremely worried about the possibly fatal effect on all living things if large numbers of such bombs were ever used in a war. He confided his worries to Bertrand Russell, whom he had met on the BBC program.

After consultations with Albert Einstein and others, Lord Russell drafted a document warning of the grave dangers presented by fission-fusion-fission bombs. The last act of Einstein's life was to sign this document, which became known as the Russell-Einstein Manifesto. It was later signed by Max Born, Fréderic Joliot-Curie, Leopold Infeld, Joseph Rotblat, Linus Pauling, Herman J. Muller, Hideki Yukawa, P.W. Bridgeman and C.F. Powell. With the exception of Infeld and Rotblat, all of them were Nobel Laureates. On July 9, 1955, with Rotblat in the chair, Russell read the Manifesto to a packed press conference.

The document contains the words: "Here then is the problem that we present to you, stark and dreadful and inescapable: Shall we put an end to the human race, or shall mankind renounce war?... There lies before us, if we choose, continual progress in happiness, knowledge and wisdom. Shall we, instead, choose death because we cannot forget our quarrels? We appeal as human beings to human beings: Remember your humanity, and forget the rest. If you can do so, the way lies open to a new Paradise; if you cannot, there lies before you the risk of universal death."

In 1945, with the horrors of World War II fresh in everyone's minds, the United Nations had been established with the purpose of eliminating war. A decade later, the Russell-Einstein Manifesto reminded the world that war *must* be abolished as an institution because of the constantly-increasing and potentially catastrophic power of modern weapons.

In 1955 the Soviets exploded their first thermonuclear device, followed in 1957 by the UK. In 1961 the USSR exploded a thermonuclear bomb with a yield of 58 megatons. A bomb of this size, four thousand times the size of the Hiroshima bomb, would be able to totally destroy a city even if it missed it by 50 kilometers. Fallout casualties would extend to a far greater distance.

In the late 1950's General Gavin, Chief of Army Research and Development in the United States, was asked by the Symington Committee, "If we got into a nuclear war and our strategic air force made an assault in force against Russia with nuclear weapons exploded in a way where the prevailing winds would carry them south-east over Russia, what would be the effect in the way of death?"

General Gavin replied: "Current planning estimates run on the order of several hundred million deaths. That would be either way depending on which way the wind blew. If the wind blew to the south-east they would be mostly in the USSR, although they would extend into the Japanese area and perhaps down into the Philippine area. If the wind blew the other way, they would extend well back into Western Europe."

Between October 16 and October 28, 1962, the Cuban Missile Crisis occurred, an incident in which the world came extremely close to a full-scale thermonuclear war. During the crisis, President Kennedy and his advisers estimated that the chance of an all-out nuclear war with Russia was 50%.

Fig. 11.3 *Albert Camus (1913-1960), Nobel Prize for Literature (1957). Shortly after the bombings of Hiroshima and Nagasaki, Camus wrote: "Our technical civilization has just reached its greatest level of savagery. We will have to choose, in the more or less near future, between collective suicide and the intelligent use of our scientific conquests."*

Recently-released documents indicate that the probability of war was even higher than Kennedy's estimate. Robert McNamara, who was Secretary of Defense at the time, wrote later, "We came within a hairbreadth of nuclear war without realizing it. ... It's no credit to us that we missed nuclear war."

In 1964 the first Chinese nuclear weapon was tested, and this was followed in 1967 by a Chinese thermonuclear bomb with a yield of 3.3 megatons. France quickly followed suit testing a fission bomb in 1966 and a thermonuclear bomb in 1968. In all about thirty nations contemplated building nuclear weapons, and many made active efforts to do so.

Fig. 11.4 *A 15 megaton bomb exploded by the United States at Bikini Atoll in the Pacific. It was more than 1000 times more powerful than the bombs that destroyed Hiroshima and Nagasaki. Fallout killed a Japanese fisherman 135 kilometers from the explosion.*

Because the concept of deterrence required an attacked nation to be able to retaliate massively even though many of its weapons might be destroyed by a preemptive strike, the production of nuclear warheads reached insane heights, driven by the collective paranoia of the Cold War. More than 50,000 nuclear warheads were deployed worldwide, a large number of them thermonuclear, and far more were manufactured. The collective explosive power of these warheads was equivalent to 20,000,000,000 tons of TNT, i.e., 4 tons for every man, woman and child on the planet, or, expressed differently, a million times the explosive power of the bomb that destroyed Hiroshima.

Flaws in the concept of nuclear deterrence

Before discussing other defects in the concept of deterrence, it must be said very clearly that the idea of "massive nuclear retaliation" is completely

Fig. 11.5 *Albert Einstein wrote: "The unleashed power of the atom has changed everything except our ways of thinking, and we thus drift toward unparalled catastrophes." He also said, "I don't know what will be used in the next world war, but the 4th will be fought with stones."*

unacceptable from an ethical point of view. The doctrine of retaliation, performed on a massive scale against civilian populations, with complete disregard for questions of guilt or innocence, violates not only the principles of common human decency and common sense, but also the the ethical principles of every major religion.

When a suspected criminal is tried for a wrongdoing, great efforts are devoted to clarifying the question of guilt or innocence. Punishment only follows if guilt can be proved beyond any reasonable doubt. Contrast this with the totally indiscriminate mass slaughter that results from a nuclear attack, the victims of which include children who could not remotely be thought of as having any degree of guilt.

It might be objected that disregard for the guilt or innocence of victims is a universal characteristic of modern war, since statistics show that, with time, a larger and larger percentage of the victims of war have been civilians, and especially children. For example, the air attacks on Coventry during World War II, or the fire bombings of Dresden and Tokyo, produced massive casualties which involved all segments of the population. The answer is that modern war has become generally unacceptable from an ethical point of view, and this unacceptability is epitomized by nuclear weapons.

The enormous and indiscriminate destruction produced by nuclear weapons formed the background for an historic 1996 decision by the International Court of Justice in the Hague. In response to questions put to it by WHO and the UN General Assembly, the Court ruled that "the threat and use of nuclear weapons would generally be contrary to the rules of international law applicable in armed conflict, and particularly the principles and rules of humanitarian law." In addition, the World Court ruled unanimously that "there exists an obligation to pursue in good faith *and bring to a conclusion* negotiations leading to nuclear disarmament in all its aspects under strict international control."

Judge Fleischhauer of Germany said in his separate opinion, "The nuclear weapon is, in many ways, the negation of the humanitarian considerations underlying the law applicable in armed conflict and the principle of neutrality. The nuclear weapon cannot distinguish between civilian and military targets. It causes immeasurable suffering. The radiation released by it is unable to respect the territorial integrity of neutral States."

President Bedjaoui, summarizing the majority opinion, called nuclear weapons "the ultimate evil", and said "By its nature, the nuclear weapon, this blind weapon, destabilizes humanitarian law, the law of discrimination in the use of weapons. ...The ultimate aim of every action in the field of nuclear arms will always be nuclear disarmament, an aim which is no longer Utopian and which all have a duty to pursue more actively than ever."

Thus the concept of nuclear deterrence is not only unacceptable from the standpoint of ethics; it is also contrary to international law. The World Court's 1996 Advisory Opinion unquestionably also represents the opinion of the majority of the world's peoples. Although no formal plebiscite has been taken, the votes in numerous resolutions of the UN General Assembly speak very clearly on this question. For example the New Agenda Resolution (53/77Y) was adopted by the General Assembly on 4 December 1998 by a massively affirmative vote, in which only 18 out of the 170 member states voted against the the resolution.[2] The New Agenda Resolution

[2]Of the 18 countries that voted against the New Agenda Resolution, 10 were Eastern

proposes numerous practical steps towards complete nuclear disarmament, and it calls on the Nuclear-Weapon States "to demonstrate an unequivocal commitment to the speedy and total elimination of their nuclear weapons and without delay to pursue in good faith and bring to a conclusion negotiations leading to the elimination of these weapons, thereby fulfilling their obligations under Article VI of the Treaty on the Non-Proliferation of Nuclear Weapons (NPT)". Thus, in addition to being ethically unacceptable and contrary to international law, nuclear weapons also contrary to the principles of democracy.

Having said said these important things, we can now turn to some of the other defects in the concept of nuclear deterrence. One important defect is that nuclear war may occur through accident or miscalculation - through technical defects or human failings. This possibility is made greater by the fact that despite the end of the Cold War, thousands of missiles carrying nuclear warheads are still kept on a "hair-trigger" state of alert with a quasi-automatic reaction time measured in minutes. There is a constant danger that a nuclear war will be triggered by error in evaluating the signal on a radar screen. For example, the BBC reported recently that a group of scientists and military leaders are worried that a small asteroid entering the earths atmosphere and exploding could trigger a nuclear war if mistaken for a missile strike.

A number of prominent political and military figures (many of whom have ample knowledge of the system of deterrence, having been part of it, have expressed concern about the danger of accidental nuclear war. Colin S. Grey[3] expressed this concern as follows: "The problem, indeed the enduring problem, is that we are resting our future upon a nuclear deterrence system concerning which we cannot tolerate even a single malfunction." General Curtis E. LeMay[4] has written, "In my opinion a general war will grow through a series of political miscalculations and accidents rather than through any deliberate attack by either side." Bruce G. Blair[5] has has remarked that "It is obvious that the rushed nature of the process, from warning to decision to action, risks causing a catastrophic mistake."... "This system is an accident waiting to happen."

"But nobody can predict that the fatal accident or unauthorized act will never happen", Fred Ikle of the Rand Corporation has written, "Given the huge and far-flung missile forces, ready to be launched from land and

European countries hoping for acceptance into NATO, whose votes seem to have been traded for increased probability of acceptance.

[3]Chairman, National Institute for Public Policy.

[4]Founder and former Commander in Chief of the United States Strategic Air Command.

[5]Brookings Institution.

sea on on both sides, the scope for disaster by accident is immense. ... In a matter of seconds - through technical accident or human failure - mutual deterrence might thus collapse."

Another serious failure of the concept of nuclear deterrence is that it does not take into account the possibility that atomic bombs may be used by terrorists. Indeed, the threat of nuclear terrorism has today become one of the most pressing dangers that the world faces, a danger that is particularly acute in the United States.

Since 1945, more than 3,000 metric tons (3,000,000 kilograms) of highly enriched uranium and plutonium have been produced - enough for several hundred thousand nuclear weapons. Of this, roughly a million kilograms are in Russia, inadequately guarded, in establishments where the technicians are poorly paid and vulnerable to the temptations of bribery. There is a continuing danger that these fissile materials will fall into the hands of terrorists, or organized criminals, or irresponsible governments. Also, an extensive black market for fissile materials, nuclear weapons components, etc., has recently been revealed in connection with the confessions of Pakistan's bomb-maker, Dr. A.Q. Khan. Furthermore, if Pakistan's less-than-stable government should be overthrown, complete nuclear weapons could fall into the hands of terrorists.

On November 3, 2003, Mohamed ElBaradei, Director General of the International Atomic Energy Agency, made a speech to the United Nations in which he called for "limiting the processing of weapons-usable material (separated plutonium and high enriched uranium) in civilian nuclear programs - as well as the production of new material through reprocessing and enrichment - by agreeing to restrict these operations to facilities exclusively under international control." It is almost incredible, considering the dangers of nuclear proliferation and nuclear terrorism, that such restrictions were not imposed long ago. Nuclear reactors used for "peaceful" purposes unfortunately also generate fissionable isotopes of plutonium, neptunium and americium. Thus all nuclear reactors must be regarded as ambiguous in function, and all must be put under strict international control. One might ask, in fact, whether globally widespread use of nuclear energy is worth the danger that it entails.

We must remember the remark of U.N. Secretary General Kofi Annan after the 9/11/2001 attacks on the World Trade Center. He said, "*This time* it was not a nuclear explosion". The meaning of his remark is clear: If the world does not take strong steps to eliminate fissionable materials and nuclear weapons, it will only be a matter of time before they will be used in terrorist attacks on major cities. Neither terrorists nor organized criminals can be deterred by the threat of nuclear retaliation, since they have no territory against which such retaliation could be directed. They blend invisibly into the general population. Nor can a "missile defense system" prevent terrorists from using nuclear weapons, since the weapons can be brought into a port in any one of the hundreds of thousands of containers that enter on ships each year, a number far too large to be checked exhaustively.

In this dangerous situation, the only logical thing for the world to do is to get rid of both fissile materials and nuclear weapons as rapidly as possible. We must acknowledge that the idea of nuclear deterrence is a dangerous fallacy, and acknowledge that the development of military systems based on nuclear weapons has been a terrible mistake, a false step that needs to be reversed. If the most prestigious of the nuclear weapons states can sincerely acknowledge their mistakes and begin to reverse them, nuclear weapons will seem less glamorous to countries like India, Pakistan, North Korea and Iran, where they now are symbols of national pride and modernism.

Future weapons

Thermonuclear weapons are genocidal and antihuman, but even worse weapons may be developed in the future through the misuse of science. A particularly repellent idea has recently been put forward: racially selective bio-weapons. Basically the idea is this: The Human Genome Project has revealed that the sequences of junk DNA (i.e., sequences that do not code for useful proteins) are racially specific. Thus the various races of humankind can be identified by looking at their junk DNA sequences. This being so, it should in principle be possible to construct a virus or toxin that will selectively attack people of a particular race.

This idea is particularly abhorrent because it simultaneously violates two important principles of human solidarity. The first principle is that, since disease is the common enemy of mankind, all humans must to work together for its eradication. The second is that all humans must regard each other as members of a single large family. This is absolutely necessary if we are to survive on our small planet.

11.3 War as a business

Eisenhower's farewell address

Because the world spends more than a trillion dollars each year on armaments, it follows that very many people make their living from war. This is the reason why it is correct to speak of war as a social, political and economic institution, and also one of the main reasons why war persists, although everyone realizes that it is the cause of much of the suffering of humanity. We know that war is madness, but it persists. We know that it threatens the survival of our species, but it persists, entrenched in the attitudes of historians, newspaper editors and television producers, entrenched in the methods by which politicians finance their campaigns, and entrenched in the financial power of arms manufacturers - entrenched also in the ponderous and costly hardware of war, the fleets of warships, bombers, tanks, nuclear missiles and so on.

In his farewell address, US President Dwight D. Eisenhower warned his nation against the excessive power that had been acquired during World War II by the military-industrial complex: "We have been compelled to create an armaments industry of vast proportions," Eisenhower said, "...Now this conjunction of an immense military establishment and a large arms industry is new in American experience. The total influence - economic, political, even spiritual - is felt in every city, every state house, every office in the federal government. ... We must not fail to comprehend its grave implications. Our toil, resources and livelihood are all involved; so is the very structure of our society. ... We must stand guard against the acquisition of unwarranted influence, whether sought or unsought, by the military-industrial complex. The potential for the disastrous rise of misplaced power exists and will persist. We must never let the weight of this combination endanger our democratic processes. We should take nothing for granted."

The devil's dynamo

Today, the term "military-industrial complex" is much too narrow. The vast and dangerous conglomerate of powerholders described by Eisenhower now includes politics, science, technology and the mass media, all feeding on the trillion dollars of global military expenditure. A circular flow of money drives this "devil's dynamo". The mass media are purchased by war-related industries, voters are influenced by the media, politicians are re-elected, and more money is poured back into the war system. The devil's dynamo needs conflicts and threats; without widely held security fears, it would wither; and hence the enemies of World War II were quickly replaced by the "menace of communism". More recently, not long after the end of the Cold War, the enemies of that conflict were replaced by the "war on terror".

A solution to the problem of war requires that voters should clearly see the augmented military-industrial complex for what it is - a threat to both peace and democracy. President Eisenhower's far-seeing farewell address deserves to be remembered and studied. It should be part of the curriculum of every school.

Resource wars

The wars of the colonial era can be seen as resource wars; and many of today's conflicts in the Middle East, Central Asia, Latin America and Africa can perhaps also be seen in this light. There is a danger that the era of resource scarcity that the world is now entering will be characterized by bitter wars for the possession of increasingly scarce oil, water and metals.

In his book, *Resource Wars: The New Landscape of Global Conflict* (2002), Michael T. Klare[6] shows that many recent wars can be interpreted as struggles for the control of natural resources. In order that such conflicts should not become more frequent in the future, a cooperative attitude towards resources is needed. The global community must face resource scarcity with solidarity.

There are many historical instances where a cooperative attitude towards resources has strengthened peaceful relationships between nations. For example, Bolivia and Peru, the two nations that share Lake Titicaca, have created a joint institution to regulate use of water from the lake and to protect it from pollution - the Autonomous Water Authority.

Another large lake shared by several countries is the Aral Sea. The lake itself is shared by Kazakhstan and Uzbekistan, but the fresh water basin of the Aral Sea also includes Afghanistan, the Kyrgyz Republic, Tajikistan and Turkmenistan. After having been reduced to half its size by massive diversion of water, the Aral Sea is now starting to refill after the completion of the Kok-Aral Dam. The countries of the Aral Sea's fresh water basin are cooperating to improve irrigation, water conservation and hydroelectric power generation.

A third example of water cooperation is the U.N.-sponsored Mekong Committee (1957) (replaced by the Mekong River Commission in 1995), which coordinates water resource development among the Lower Basin nations, Cambodia, Laos, Thailand, and Vietnam. Today, the Mekong River Commission aids in the management and preservation of resources in the basin, and addresses such issues as population growth, environmental preservation and regional security.

Globally there are today more than 3,800 declarations or conventions on water, of which 286 are treaties. They demonstrate that cooperation in the field of resource management can make a valuable contribution towards building a system of international law.

---

[6]Michael Klare is the Five College Professor of Peace and World Security Studies, based at Hampshire College in Amherst Massachusetts, but also lecturing at Amherst, Mount Holyoke, Smith and the University of Massachusetts.

11.4 War as a hindrance to global equality

Indirect costs of war

The costs of war, both direct and indirect, are so enormous that they are almost beyond comprehension. Globally, the institution of war interferes seriously with the use of tax money for constructive and peaceful purposes. Today, despite the end of the Cold War, the world spends roughly a trillion (i.e., a million million) US dollars each year on armaments. This colossal flood of money could have been used instead for education, famine relief, development of infrastructure, or on urgently needed public health measures.

The World Health Organization lacks funds to carry through an anti-malarial program on as large a scale as would be desirable, but the entire program could be financed for less than our military establishments spend in a single day. Five hours of world arms spending is equivalent to the total cost of the 20-year WHO campaign that resulted in the eradication of smallpox. For every 100,000 people in the world, there are 556 soldiers, but only 85 doctors. Every soldier costs an average of $20,000 per year, while the average spent on education is only $380 per school-aged child. With a diversion of funds consumed by three weeks of military spending, the world could create a sanitary water supply for all its people, thus eliminating the cause of almost half of all human illness.

A new drug-resistant form of tuberculosis has recently become widespread in Asia and in the former Soviet Union. In order to combat this new and highly dangerous form of tuberculosis and to prevent its spread, WHO needs $500 million, an amount equivalent to 4 hours of world arms spending.

Today's world is one in which roughly ten million children die every year from starvation or from diseases related to poverty. Besides this enormous waste of young lives through malnutrition and preventable disease, there is a huge waste of opportunities through inadequate education. The rate of illiteracy in the 25 least developed countries is 80%, and the total number of illiterates in the world is estimated to be 800 million. Meanwhile every 60 seconds the world spends $2 million on armaments.

It is plain that if the almost unbelievable sums now wasted on the institution of war were used constructively, most of the pressing problems of humanity could be solved, but today the world spends more than 20 times as much on war as it does on development.

In the previous chapter we mentioned the intolerable degree of economic inequality that characterizes today's world. Development is blocked by endemic disease and by excessive population growth. If the World Health

Organization had sufficient funds to provide primary health care centers for all, universal vaccination programs, and the information and materials needed for family planning, the development-blocking problems of population growth and disease could be overcome. The money required is a tiny fraction of the vast sums now wasted on war.

Destruction of infrastructure

Most insurance policies have clauses written in fine print exempting companies from payment of damage caused by war. The reason for this is simple. The damage caused by war is so enormous that insurance companies could never come near to paying for it without going bankrupt.

We mentioned above that the world spends roughly a trillion dollars each year on preparations for war. A similarly colossal amount is needed to repair the damage to infrastructure caused by war. Sometimes this damage is unintended, but sometimes it is intentional. During World War II, one of the main aims of air attacks by both sides was to destroy the industrial infrastructure of the opponent. This made some sense in a war expected to last several years, because the aim was to prevent the enemy from producing more munitions. However, during the Gulf War of 1990, the infrastructure of Iraq was attacked, even though the war was expected to be short. Electrical generating plants and water purification facilities were deliberately destroyed with the apparent aim of obtaining leverage over Iraq after the war.

In general, because war has such a catastrophic effect on infrastructure, it can be thought of as the opposite of development. War is the greatest generator of poverty.

Environmental damage

Warfare during the 20th century has not only caused the loss of 175 million lives (primarily civilians) - it has also caused the greatest ecological catastrophes in human history. The damage takes place even in times of peace. Studies by Joni Seager, a geographer at the University of Vermont, conclude that "a military presence anywhere in the world is the single most reliable predictor of ecological damage".

Modern warfare destroys environments to such a degree that it has been described as an "environmental holocaust." For example, herbicides use in the Vietnam War killed an estimated 6.2 billion board-feet of hardwood trees in the forests north and west of Saigon, according to the American Association for the Advancement of Science. Herbicides such as Agent Orange also made enormous areas of previously fertile land unsuitable for agriculture for many years to come[7]. In Vietnam and elsewhere in the world, valuable agricultural land has also been lost because land mines or the remains of cluster bombs make it too dangerous for farming.

During the Gulf War of 1990, the oil spills amounted to 150 million barrels, 650 times the amount released into the environment by the notorious Exxon Valdez disaster. During the Gulf War an enormous number of shells made of depleted uranium were fired. When the dust produced by exploded shells is inhaled it often produces cancer, and it will remain in the environment of Iraq for decades.

Radioactive fallout from nuclear tests pollutes the global environment and causes many thousands of cases of cancer, as well as birth abnormalities. Most nuclear tests have been carried out on lands belonging to indigenous peoples.

11.5 Global inequalities as a hindrance to peace

Inequalities maintained by force

Global economic inequalities are linked to the war system in many ways. We have just discussed how the enormous amounts of money wasted on war reduce the funds available for population stabilization and development, and how war destroys infrastructure and the environment. Another link can be found in the fact that powerful industrialized countries maintain unequal economic relationships with developing countries by means of military force, or through political pressures which, in the last analysis, rely on force. This was very obvious during the colonial era, when the use of force for economic reasons was exemplified by the Boar War (1899-1902), Belgian military acquisition of the Congo, the French army in North Africa, the British army in India, and by "gunboat diplomacy" throughout the world. The recent resource wars described by Michael Klare show that economic motivations lie behind much of the warfare that can be observed today.

Rich nations fear global democracy

An indirect relationship between global economic inequalities and war can be found in the fact that the enormous contrasts between rich and poor block the development of global governance. There are many reasons why effective governance at the global level is needed. In a world of rapidly in-

[7]Agent Orange also produced cancer, birth abnormalities and other serious forms of illness both in the Vietnamese population and among the foreign soldiers fighting in Vietnam.

creasing interdependence, a world where modern weapons have made war prohibitively dangerous, an effective system of international law and governance is urgently needed. But rich nations will not allow these things, because they fear that they will lose their privileged positions.

The recent development of the European Federation is extremely interesting because it might serve as a model for the development and reform of the United Nations. In the formation of the European Union and its extension to the east, there were worries about contrasts between rich and poor regions. Would the wealthy nations of Europe be excessively taxed to help their poorer neighbors? Would workers from the poorer parts of Europe migrate in excessive numbers to the richer regions? In the case of the EU, as in the reunification of Germany, serious problems were present because of economic inequality. But the necessary sacrifices and adjustments were made, and the problems were overcome. The motives for unification were strong: Europe had been a central battleground in two world wars, and the statesmen of Europe were determined that such tragedies should never be repeated. One can hope that global inequalities can similarly be overcome by a world motivated by the desire for peace, justice and stability.

11.6 The future of global governance

Peace within nations

The problem of achieving internal peace over a large geographical area is not insoluble. It has already been solved. There exist today many nations or regions within each of which there is internal peace, and some of these are so large that they are almost worlds in themselves. One thinks of China, India, Brazil, Australia, the Russian Federation, the United States, and the European Union. It must of course be acknowledged that the large political units just named are neither perfectly just nor perfectly peaceful. Nevertheless, they all have achieved a high degree of internal peace and political cohesion. Many of these enormous societies contain a variety of ethnic groups, a variety of religions and a variety of languages, as well as striking contrasts between wealth and poverty. If these great land areas have been forged into peaceful and cooperative societies, cannot the same methods of government be applied globally?

But what are the methods that nations use to achieve internal peace? Firstly, every true government needs to have the power to make and enforce laws that are binding on individual citizens. Secondly the power of taxation is a necessity. These are two requirements of every true government; but there is a third point that still remains to be discussed:

Within their own territories, almost all nations have more military power than any of their subunits. For example, the US Army is more powerful than the State Militia of Illinois. This unbalance of power contributes to the stability of the Federal Government of the United States. When the FBI wanted to arrest Al Capone, it did not have to bomb Chicago. Agents just went into the city and arrested the gangster. Even if Capone had been enormously popular in Illinois, the the government of the state would have realized in advance that it had no chance of resisting the US Federal Government, and it still would have allowed the "Feds" to make their arrest. Similar considerations hold for almost all nations within which there is internal peace. It is true that there are some nations within which subnational groups have more power than the national government, but these are frequently characterized by civil wars.

Of the large land areas within which internal peace has been achieved, the European Union differs from the others because its member states still maintain powerful armies. The EU forms a realistic model for what can be achieved globally in the near future by reforming and strengthening the United Nations. In the distant future, however, we can imagine a time when a world federal authority will have much more power than any of its member states, and when national armies will have only the size needed to maintain local order.

Today there is a pressing need to enlarge the size of the political unit from the nation-state to the entire world. The need to do so results partly from the terrible dangers of modern weapons and partly from global economic interdependence. The progress of science has created this need, but science has also given us the means to enlarge the political unit: Our almost miraculous modern communications media, if properly used, have the power to weld all of humankind into a single supportive and cooperative society.

Federations

A federation of states is, by definition, a limited union where the federal government has the power to make laws that are binding on individuals, but where the laws are confined to interstate matters, and where all powers not expressly delegated to the federal government are retained by the individual states. In other words, in a federation each of the member states runs its own internal affairs according to its own laws and customs, but in certain agreed-on matters, where the interests of the states overlap, authority is specifically delegated to the federal government.

For example, if the nations of the world considered the control of narcotics to be a matter of mutual concern; if they agreed to set up a commission with the power to make laws preventing the growing, refinement and distribution of harmful drugs, and the power to arrest individuals for violating those laws, then we would have a world federation in the area of narcotics control.

If the community of nations decided to give the federal authority the additional power to make laws defining the rights and obligations of multinational corporations, and the power to arrest or fine individuals violating those laws, then we would have a world federation with even broader powers; but these powers would still be carefully defined and limited.

In setting up a federation, the member states can decide which powers they wish to delegate to it; and all powers not expressly delegated are retained by the individual states. We are faced with the problem of constructing a new world order which will preserve the advantages of local self-government while granting certain carefully-chosen powers to larger regional or global authorities. Which things should be decided locally or regionally and which globally?

In the future, global overpopulation and famine will become increasingly difficult and painful problems. Since various cultures take widely different attitudes towards birth control and family size, the problem of overpopulation seems to be one which which should be solved locally or regionally; and no country or region should be allowed to export its population problem by sending large numbers of its citizens abroad. By contrast, global pollution as well as security and controls on the manufacture and export of armaments will require an effective authority at the global level. It should also be the responsibility of the international community to intervene to prevent gross violations of human rights.

Since the federal structure seems well suited to a world government with limited and carefully-defined powers that would preserve as much local autonomy as possible, it is worthwhile to look at the histories of a few federations. For example, we can learn much by looking at the history of the federal government of the United States.

George Mason, one of the architects of the federal constitution of the United States, believed that "such a government was necessary as could directly operate on individuals, and would punish those only whose guilt required it," while James Madison, another drafter of the U.S. federal constitution, remarked that the more he reflected on the use of force, the more he doubted "the practicability, the justice and the efficacy of it when applied to people collectively, and not individually." Finally, Alexander Hamilton, in his Federalist Papers, discussed the Articles of Confederation with the following words: "To coerce the states is one of the maddest projects that was ever devised. ... Can any reasonable man be well disposed towards a government which makes war and carnage the only means of supporting itself - a government that can exist only by the sword? Every such war must involve the innocent with the guilty. The single consideration should be enough to dispose every peaceable citizen against such a government. ... What is the cure for this great evil? Nothing, but to enable the... laws to operate on individuals, in the same manner as those of states do."

The United Nations has a charter analogous to the Articles of Confederation, which preceded the U.S. Federal Constitution. The Articles of Confederation proved to be too weak, just as the present structure of the U.N. is proving to be too weak. It acts by attempting to coerce states, a procedure which Alexander Hamilton characterized as "one of the maddest projects that was ever devised." Whether this coercion takes the form of economic sanctions, or whether it takes the form of military intervention, the practicability, the justice and the efficacy of the U.N.'s efforts are hampered because they are applied to people collectively and not one by one. What is the cure for this great evil? "Nothing", Hamilton tells us, "but to enable the laws to act on individuals, in the same manner as those of states do."

Laws binding on individuals

In 1998, in Rome, representatives of 120 countries signed a statute establishing a International Criminal Court, with jurisdiction over the crime of genocide, crimes against humanity, war crimes, and the crime of aggression. Four years were to pass before the necessary ratifications were gathered, but by Thursday, April 11, 2002, 66 nations had ratified the Rome agreement - 6 more than the 60 needed to make the court permanent.

It would be impossible to overstate the importance of the International Criminal Court. At last international law acting on individuals has become a reality! The only effective and just way that international laws can act is to make individuals responsible and punishable, since (in the words of Alexander Hamilton), "To coerce states is one of the maddest projects ever devised." In an increasingly interdependent world, international law has become a necessity. We cannot have peace and justice without it. But the coercion of states is neither just[8] nor feasible, and therefore international laws must act on individuals.

The jurisdiction of the ICC is at present limited to a very narrow class of crimes. In fact, the ICC does not at present act on the crime of aggression, although this crime is listed in the Rome Statute, and although there are plans for its future inclusion in the ICC's activities. The global community will have a chance to see how the court works in practice, and

[8]Because it punishes the innocent as well as the guilty.

in the future the community will undoubtedly decide to broaden the ICC's range of jurisdiction.

The Tobin Tax

A strengthened UN would need a reliable source of income to make the organization less dependent on wealthy countries, which tend to give support only to those interventions of which they approve. A promising solution to this problem is the so-called "Tobin tax", named after the Nobel-laureate economist James Tobin of Yale University. Tobin proposed that international currency exchanges should be taxed at a rate between 0.1 and 0.25 percent. He believed that even this extremely low rate of taxation would have the beneficial effect of damping speculative transactions, thus stabilizing the rates of exchange between currencies. When asked what should be done with the proceeds of the tax, Tobin said, almost as an afterthought, "Let the United Nations have it."

The volume of money involved in international currency transactions is so enormous that even the tiny tax proposed by Tobin would provide the United Nations with between 100 billion and 300 billion dollars annually. By strengthening the activities of various UN agencies, such as WHO, UNESCO and FAO, the additional income would add to the prestige of the United Nations and thus make the organization more effective when it is called upon to resolve international political conflicts.

Besides the Tobin tax, other measure have been proposed to increase the income of the United Nations. For example, it has been proposed that income from resources of the sea bed be given to the UN, and that the UN be given the power to tax carbon dioxide emissions. All of the proposals for giving the United Nations an adequate income have been strongly opposed by a few nations that wish to control the UN through its purse strings. However, it is absolutely essential for the future development of the United Nations that the organization be given the power to impose taxes. No true government can exist without this power. It is just as essential as is the power to make and enforce laws that are binding on individuals.

Voting reforms

A serious weakness of the present United Nations Charter is the principle of "one nation one vote" in the General Assembly. This principle seems to establish equality between nations, but in fact it is very unfair: For example it gives a citizen of China or India less than a thousandth the voting power of a citizen of Malta or Iceland. A reform of the voting system is clearly needed.

Among the proposals for reform is the idea of having final votes cast by blocks. For example, Europe could form a block, Africa another, and so on. A second proposal is that the General Assembly might be supplemented by a People's Assembly.

The veto right in the Security Council is clearly a fault in the present structure of the U.N. It has been suggested that the rules should be changed so that a veto in the Security Council could be over-ruled by a two thirds majority vote of the General Assembly. Other reform proposals call for the abolition of the veto in the Security Council, or even for the abolition of the Security Council itself.

11.7 Global ethics

Education for world citizenship

Besides a humane, democratic and just framework of international law and governance, we urgently need a new global ethic, - an ethic where loyalty to family, community and nation will be supplemented by a strong sense of the brotherhood of all humans, regardless of race, religion or nationality. Schiller expressed this feeling in his "Ode to Joy", a part of which is the text of Beethoven's Ninth Symphony. Hearing Beethoven's music and Schiller's words, most of us experience an emotion of resonance and unity with the message: All humans are brothers and sisters - not just some - all! It is almost a national anthem of humanity. The feelings that the music and words provoke are similar to patriotism, but broader. It is this higher loyalty to humanity as a whole, this sense of a universal human family, that we need to cultivate in education, in the mass media, and in religion.

Educational reforms are urgently needed, particularly in the teaching of history. As it is taught today, history is a chronicle of power struggles and war, told from a biased national standpoint. Our own race or religion is superior; our own country is always heroic and in the right.

We urgently need to replace this indoctrination in chauvinism by a reformed view of history, where the slow development of human culture is described, giving adequate credit to all who have contributed. Our modern civilization is built on the achievements of many ancient cultures. China, Japan, India, Mesopotamia, Egypt, Greece, the Islamic world, Christian Europe, and the Jewish intellectual traditions all have contributed. Potatoes, corn, squash, vanilla, chocolate, chili peppers, pineapples, quinine, etc. are gifts from the American Indians. Human culture, gradually built up over thousands of years by the patient work of millions of hands and minds, should be presented as a precious heritage - far too precious to be risked in a thermonuclear war.

Reform is also urgently needed in the teaching of economics and business. The economics of growth must be replaced by equilibrium economics, where considerations of ecology, carrying capacity, and sustainability are given their proper weight, and where the quality of life of future generations has as much importance as present profits.

Secondly, the education of economists and businessmen needs to face the problems of global poverty - the painful contrast between the affluence and wastefulness of the industrial North and the malnutrition, disease and illiteracy endemic in the South. Students of economics and business must look for the roots of poverty not only in population growth and war, but also in the history of colonialism and neocolonialism, and in defects in global financial institutions and trade agreements. They must be encouraged to formulate proposals for the correction of North-South economic inequality.

The economic impact of war and preparation for war should be included in the training of economists. Both the direct and indirect costs of war should be studied, for example the effect of unimaginably enormous military budgets in reducing the money available to solve pressing problems posed by the resurgence of infectious disease (e.g. AIDS, and drug-resistant forms of malaria and tuberculosis); the problem of population stabilization; food problems; loss of arable land; future energy problems; the problem of finding substitutes for vanishing nonrenewable resources, and so on.

Finally, economics curricula should include the problems of converting war-related industries to peaceful ones - the problem of beating swords into plowshares. It is often said that our economies are dependent on arms industries. If this is so, it is an unhealthy dependence, analogous to drug addiction, since arms industries do not contribute to future-oriented infrastructure. The problem of conversion is an important one. It is the economic analog of the problem of ending a narcotics addiction, and it ought to be given proper weight in the education of economists.

Law students should be made aware of the importance of international law. They should be familiar with its history, starting with Grotius and the Law of the Sea. They should know the histories of the International Court of Justice and the Nüremberg Principles. They should study the United Nations Charter (especially the articles making war illegal) and the Universal Declaration of Human Rights, as well as the Rome Treaty and the foundation of the International Criminal Court. They should be made aware of a deficiency in the present United Nations - the lack of a legislature with the power to make laws that are binding on individuals.

Students of law should be familiar with all of the details of the World Court's historic Advisory Opinion on Nuclear Weapons, a decision that make the use or threat of use of nuclear weapons illegal. They should also study the Hague and Geneva Conventions, and the various international treaties related to nuclear, chemical and biological weapons. The relationship between the laws of the European Union and those of its member states should be given high importance. The decision by the British Parliament that the laws of the EU take precedence over British law should be a part of the curriculum.

In teaching science too, reforms are needed. Graduates in science and engineering should be conscious of their responsibilities. They must resolve never to use their education in the service of war, nor for the production of weapons, nor in any way that might be harmful to society or to the environment.

Science and engineering students ought to have some knowledge of the history and social impact of science. They could be given a course on the history of scientific ideas, and in connection with modern historical developments such as the industrial revolution, the global population explosion, the development of nuclear weapons, genetic engineering, and information technology, some discussion of social impact could be introduced. One might hope to build up in science and engineering students an understanding of the way in which their own work is related to the general welfare of humankind, and a sense of individual social and ethical responsibility. These elements are needed in science education if rapid technological progress is to be beneficial to society rather than harmful.

The role of the mass media

In the mid-1950's, television became cheap enough so that ordinary people in the industrialized countries could afford to own sets. During the infancy of television, its power was underestimated. The great power of television is due to the fact that it grips two senses simultaneously, both vision and hearing. The viewer becomes an almost-hypnotized captive of the broadcast. In the 1950's, this enormous power, which can be used both for good and for ill, was not yet fully apparent. Thus insufficient attention was given to the role of television in education, in setting norms, and in establishing values. Television was not seen as an integral part of the total educational system.

Although the intergenerational transmission of values, norms, and culture is much less important in industrial societies than it is in traditional ones, modern young people of the west and north are by no means at a loss over where to find their values, fashions and role models. With every breath they inhale the values and norms of the mass media. Totally surrounded by a world of television and film images, they accept this world as their own. Unfortunately the culture of television, films and computer games is more often a culture of violence than a culture of peace.

Computer games designed for young boys often give the strongest imaginable support to our present culture of violence. For example, a game entitled "Full Spectrum Warrior" was recently reviewed in a Danish newspaper. According to the reviewer, "...An almost perfect combination of graphics, sound, band design, and gameplay makes it seem exactly like the film Black Hawk Down - with the player as the main character. This is not just a coincidence, because the game is based on an army training program. ... Full Spectrum Warrior is an extremely intense experience, and despite the advanced possibilities, the controls are simple enough so that young children can play it. ... The player is completely drawn into the screen, and remains there until the end of the mission." The reviewer gave the game six stars (the maximum).

If entertainment is evaluated only on the basis of popularity, what might be called "the pornography of violence" gets high marks. However, there is another way of looking at entertainment. It is a part, and a very important part, of our total educational system. In modern industrial societies, this important educational function has been given by default to commercial interests. We would not want Coca Cola to run our schools, but entertainment is just as important as the school or home environment in forming values and norms, and entertainment is in the hands of commerce.

Today we are faced with the task of creating a new global ethic in which loyalty to family, religion and nation will be supplemented by a higher loyalty to humanity as a whole. In addition, our present culture of violence must be replaced by a culture of peace. To achieve these essential goals, we urgently need the cooperation of the mass media.

One is faced with a dilemma, because on the one hand artistic freedom is desirable and censorship undesirable, but on the other hand some degree of responsibility ought to be exercised by the mass media because of their enormous influence in creating norms and values.

Of course we cannot say to the entertainment industry, "From now on you must not show anything but David Attenborough and the life of Gandhi". However, it would be enormously helpful if every film or broadcast or computer game could be evaluated not only for its popularity and artistic merit, but also in terms of the good or harm that it does in the task of building a peaceful world.

Why doesn't the United Nations have its own global television and radio network? Such a network could produce an unbiased version of the news. It could broadcast documentary programs on global problems. It could produce programs showing viewers the music, art and literature of other cultures than their own. It could broadcast programs on the history of ideas, in which the contributions of many societies were adequately recognized. At

New Year, when people are in the mood to think of the past and the future, the Secretary General of the United Nations could broadcast a "State of the World" message, summarizing the events of the past year and looking forward to the new year, with its problems, and with his recommendations for their solution. A United Nations television and radio network would at least give viewers and listeners a choice between programs supporting militarism, and programs supporting a global culture of peace. At present they have little choice.

The role of religion

Finally, let us turn to religion, with its enormous influence on human thought and behavior.

In the 6th century B.C., Prince Gautama Buddha founded a new religion in India, with a universal (non-tribal) code of ethics. Among the sayings of the Buddha are as follows:

"Hatred does not cease by hatred at any time; hatred ceases by love."

"Let a man overcome anger by love; let him overcome evil by good."

"All men tremble at punishment. All men love life. Remember that you are like them, and do not cause slaughter."

Similarly, Christianity offers a strongly-stated ethic, which, if practiced, would make war impossible. In Mathew, the following passage occurs:

"Ye have heard it said: Thou shalt love thy neighbor and hate thy enemy. But I say unto you: Love your enemies, bless them that curse you, do good to them that hate you, and pray for them that spitefully use you and persecute you."

This seemingly impractical advice - that we should love our enemies - is in fact of the greatest practicality, since acts of unilateral kindness and generosity can stop escalatory cycles of revenge and counter-revenge such as those that characterize the present conflicts in the Middle East and the recent troubles in Northern Ireland. However, Christian nations, while claiming to adhere to the ethic of love and forgiveness, have adopted a policy of "massive retaliation". involving systems of thermonuclear missiles whose purpose is to destroy as much as possible of the country at which the retaliation is aimed. It is planned that whole populations should be killed in a "massive retaliation", innocent children along with guilty politicians.

The startling contradiction between what Christian nations profess and what they do was obvious even before the advent of nuclear weapons, at the time when Leo Tolstoy, during his last years, was exchanging letters with a young Indian lawyer in South Africa. In one of his letters to Gandhi, Tolstoy wrote:

"...The longer I live, and especially now, when I vividly feel the nearness of death, the more I want to tell others what I feel so particularly clearly and what to my mind is of great importance - namely that which is called passive resistance, but which is in reality nothing else but the teaching of love, uncorrupted by false interpretations. That love - i.e. the striving for the union of human souls and the activity derived from that striving - is the highest and only law of human life, and in the depth of his soul every human being knows this (as we most clearly see in children); he knows this until he is entangled in the false teachings of the world. This law was proclaimed by all - by the Indian as by the Chinese, Hebrew, Greek and Roman sages of the world. I think that this law was most clearly expressed by Christ, who plainly said that 'in this alone is all the law and the prophets.' ..."

"...The peoples of the Christian world have solemnly accepted this law, while at the same time they have permitted violence and built their lives on violence; and that is why the whole life of the Christian peoples is a continuous contradiction between what they profess, and the principles on which they order their lives - a contradiction between love accepted as the law of life, and violence which is recognized and praised, acknowledged even as a necessity..."

As everyone knows, Gandhi successfully applied the principle of non-violence to the civil rights struggle in South Africa, and later to the political movement which gave India its freedom and independence. Later, non-violence was successfully applied by Martin Luther King, and by Nelson Mandela. Gandhi was firm in pointing out that the ends do not justify the means, since violent methods inevitably contaminate the result achieved. The same theme can be seen in the following quotation from Martin Luther King.

"Why should we love our enemies?", Dr. King wrote, "Returning hate for hate multiplies hate, adding deeper darkness to a night already devoid of stars. Darkness cannot drive out darkness; only light can do that. Hate cannot drive out hate. Only love can do that. ... Love is the only force capable of transforming an enemy into a friend. We never get rid of an enemy by meeting hate with hate; we get rid of an enemy by getting rid of enmity. ... It is this attitude that made it possible for Lincoln to speak a kind word about the South during the Civil War, when feeling was most bitter. Asked by a shocked bystander how he could do this, Lincoln said, 'Madam, do I not destroy my enemies when I make them my friends?' This is the power of redemptive love."

In 1967, a year before his assassination, Dr. King forcefully condemned the Viet Nam war in an address at a massive peace rally in New York City. He felt that opposition to war followed naturally from his advocacy of non-violence. Regarding nuclear weapons, Dr. King wrote, "Wisdom born of experience should tell us that war is obsolete. There may have been a time when war served a negative good by preventing the spread of an evil force, but the power of modern weapons eliminates even the possibility that war may serve as a negative good. If we assume that life is worth living, and that man has a right to survival, then we must find an alternative to war. ... I am convinced that the Church cannot be silent while mankind faces the threat of nuclear annihilation. If the church is true to her mission, she must call for an end to the nuclear arms race."

Suggestions for further reading

(1) R. Axelrod, *The Evolution of Cooperation*, Basic Books, New York, (1984).
(2) W. Brandt, *World Armament and World Hunger: A Call for Action*, Victor Gollanz Ltd., London, (1982).
(3) E. Chivian, and others (eds.), *Last Aid: The Medical Dimensions of Nuclear War*, W.H. Freeman, San Fransisco, (1982).
(4) I. Eibl-Eibesfeldt, *The Biology of War and Peace*, Thames and Hudson, New York, (1979).
(5) R.A. Hinde, *Biological Basis for Human Social Behaviour*, McGraw-Hill, New York, (1977).
(6) R.A. Hinde, *Towards Understanding Relationships*, Academic Press, London, (1979).
(7) M. Khanert and others (eds.), *Children and War*, Peace Union of Finland, Helsinki, (1983).
(8) K. Lorentz, *On Aggression*, Bantam Books, New York, (1977).
(9) Medical Association's Board of Science and Education, *The Medical Effects of Nuclear War*, Wiley, (1983).
(10) M. Renner, *Swords into Plowshares: Converting to a Peace Economy*, Worldwatch Paper 96, Worldwatch Institute, Washington D.C., (1990).
(11) J. Rotblat (ed.), *Shaping Our Common Future: Dangers and Opportunities (Proceedings of the Forty-Second Pugwash Conference on Science and World Affairs)*, World Scientific, London, (1994).
(12) R.L. Sivard, *World Military and Social Expenditures*, World Priorities, Box 25140, Washington, D.C. 20007, (published annually).
(13) J.E. Slater, *Governance*, Aspen Institute for Humanistic Studies, New York, (1976).
(14) P.B. Smith, J.D. Schilling and A.P. Haines, *Introduction and Summary*, in *Draft Report of the Pugwash Study Group: The World at the Crossroads*, Berlin, (1992).
(15) A. Szent-Györgyi, *The Crazy Ape*, Philosophical Library, New York, (1970).

(16) J. Tinbergen (coordinator), *Reshaping the International Order*, Dutton, New York, (1976).
(17) C. Zahn-Waxler, *Altruism and Aggression: Biological and Social Origins*, Cambridge University Press, (1986).
(18) J.L. Henderson, *Hiroshima*, Longmans (1974).
(19) A. Osada, *Children of the A-Bomb, The Testament of Boys and Girls of Hiroshima*, Putnam, New York (1963).
(20) M. Hachiya, M.D., *Hiroshima Diary*, The University of North Carolina Press, Chapel Hill, N.C. (1955).
(21) M. Yass, *Hiroshima*, G.P. Putnam's Sons, New York (1972).
(22) R. Jungk, *Children of the Ashes*, Harcourt, Brace and World (1961).
(23) B. Hirschfield, *A Cloud Over Hiroshima*, Baily Brothers and Swinfin Ltd. (1974).
(24) J. Hersey, *Hiroshima*, Penguin Books Ltd. (1975).
(25) R. Rhodes, *Dark Sun: The Making of the Hydrogen Bomb*, Simon and Schuster, New York, (1995)
(26) R. Rhodes, *The Making of the Atomic Bomb*, Simon and Schuster, New York, (1988).
(27) D.V. Babst et al., *Accidental Nuclear War: The Growing Peril*, Peace Research Institute, Dundas, Ontario, (1984).
(28) S. Britten, *The Invisible Event: An Assessment of the Risk of Accidental or Unauthorized Detonation of Nuclear Weapons and of War by Miscalculation*, Menard Press, London, (1983).
(29) M. Dando and P. Rogers, *The Death of Deterrence*, CND Publications, London, (1984).
(30) N.F. Dixon, *On the Psychology of Military Incompetence*, Futura, London, (1976).
(31) D. Frei and C. Catrina, *Risks of Unintentional Nuclear War*, United Nations, Geneva, (1982).
(32) H. L'Etang, *Fit to Lead?*, Heinemann Medical, London, (1980).
(33) SPANW, *Nuclear War by Mistake - Inevitable or Preventable?*, Swedish Physicians Against Nuclear War, Lulea, (1985).
(34) J. Goldblat, *Nuclear Non-proliferation: The Why and the Wherefore*, (SIPRI Publications), Taylor and Francis, (1985).
(35) IAEA, *International Safeguards and the Non-proliferation of Nuclear Weapons*, International Atomic Energy Agency, Vienna, (1985).
(36) J. Schear, ed., *Nuclear Weapons Proliferation and Nuclear Risk*, Gower, London, (1984).
(37) D.P. Barash and J.E. Lipton, *Stop Nuclear War! A Handbook*, Grove Press, New York, (1982).
(38) C.F. Barnaby and G.P. Thomas, eds., *The Nuclear Arms Race: Control or Catastrophe*, Francis Pinter, London, (1982).

(39) L.R. Beres, *Apocalypse: Nuclear Catastrophe in World Politics*, Chicago University press, Chicago, IL, (1980).
(40) F. Blackaby et al., eds., *No-first-use*, Taylor and Francis, London, (1984).
(41) NS, ed., *New Statesman Papers on Destruction and Disarmament* (NS Report No. 3), New Statesman, London, (1981).
(42) H. Caldicot, *Missile Envy: The Arms Race and Nuclear War*, William Morrow, New York, (1984).
(43) R. Ehrlich, *Waging the Peace: The Technology and Politics of Nuclear Weapons*, State University of New York Press, Albany, NY, (1985).
(44) W. Epstein, *The Prevention of Nuclear War: A United Nations Perspective*, Gunn and Hain, Cambridge, MA, (1984).
(45) W. Epstein and T. Toyoda, eds., *A New Design for Nuclear Disarmament*, Spokesman, Nottingham, (1975).
(46) G.F. Kennan, *The Nuclear Delusion*, Pantheon, New York, (1983).
(47) R.J. Lifton and R. Falk, *Indefensible Weapons: The Political and Psychological Case Against Nuclearism*, Basic Books, New York, (1982).
(48) J.R. Macy, *Despair and Personal Power in the Nuclear Age*, New Society Publishers, Philadelphia, PA, (1983).
(49) A.S. Miller et al., eds., *Nuclear Weapons and Law*, Greenwood Press, Westport, CT, (1984).
(50) MIT Coalition on Disarmament, eds., *The Nuclear Almanac: Confronting the Atom in War and Peace*, Addison-Wesley, Reading, MA, (1984).
(51) UN, *Nuclear Weapons: Report of the Secretary-General of the United Nations*, United Nations, New York, (1980).
(52) IC, *Proceedings of the Conference on Understanding Nuclear War*, Imperial College, London, (1980).
(53) B. Russell, *Common Sense and Nuclear Warfare*, Allen and Unwin, London, (1959).
(54) F. Barnaby, *The Nuclear Age*, Almqvist and Wiksell, Stockholm, (1974).
(55) D. Albright, F. Berkhout and W. Walker, *Plutonium and Highly Enriched Uranium 1996: World Inventories, Capabilities and Policies*, Oxford University Press, Oxford, (1997).
(56) G.T. Allison et al., *Avoiding Nuclear Anarchy: Containing the Threat of Loose Russian Nuclear Weapons and Fissile Material*, MIT Press, Cambridge MA, (1996).
(57) B. Bailin, *The Making of the Indian Atomic Bomb: Science, Secrecy, and the Post-colonial State*, Zed Books, London, (1998).
(58) G.K. Bertsch and S.R. Grillot, (Eds.), *Arms on the Market: Reducing*

the Risks of Proliferation in the Former Soviet Union, Routledge, New York, (1998).
(59) P. Bidawi and A. Vanaik, *South Asia on a Short Fuse: Nuclear Politics and the Future of Global Disarmament*, Oxford University Press, Oxford, (2001).
(60) F.A. Boyle, *The Criminality of Nuclear Deterrence: Could the U.S. War on Terrorism Go Nuclear?*, Clarity Press, Atlanta GA, (2002).
(61) G. Burns, *The Atomic Papers: A Citizen's Guide to Selected Books and Articles on the Bomb, the Arms Race, Nuclear Power, the Peace Movement, and Related Issues*, Scarecrow Press, Metuchen NJ, (1984).
(62) L. Butler, *A Voice of Reason*, The Bulletin of Atomic Scientists, **54**, 58-61, (1998).
(63) R. Butler, *Fatal Choice: Nuclear Weapons and the Illusion of Missile Defense*, Westview Press, Boulder CO, (2001).
(64) R.P. Carlisle (Ed.), *Encyclopedia of the Atomic Age*, Facts on File, New York, (2001).
(65) G.A. Cheney, *Nuclear Proliferation: The Problems and Possibilities*, Franklin Watts, New York, (1999).
(66) A. Cohen, *Israel and the Bomb*, Colombia University Press, New York, (1998).
(67) S.J. Diehl and J.C. Moltz, *Nuclear Weapons and Nonproliferation: A Reference Handbook*, ABC-Clio Information Services, Santa Barbara CA, (2002).
(68) H.A. Feiveson (Ed.), *The Nuclear Turning Point: A Blueprint for Deep Cuts and De-Alerting of Nuclear Weapons*, Brookings Institution Press, Washington D.C., (1999).
(69) R. Forsberg et al., *Nonproliferation Primer: Preventing the Spread of Nuclear, Chemical and Biological Weapons*, MIT Press, Cambridge, (1995).
(70) R. Hilsman, *From Nuclear Military Strategy to a World Without War: A History and a Proposal*, Praeger Publishers, Westport, (1999).
(71) International Physicians for the Prevention of Nuclear War and The Institute for Energy and Environmental Research *Plutonium: Deadly Gold of the Nuclear Age*, International Physicians Press, Cambridge MA, (1992).
(72) R.W. Jones and M.G. McDonough, *Tracking Nuclear Proliferation: A Guide in Maps and Charts, 1998*, The Carnegie Endowment for International Peace, Washington D.C., (1998).
(73) R.J. Lifton and R. Falk, *Indefensible Weapons: The Political and Psychological Case Against Nuclearism*, Basic Books, New York, (1982).
(74) J. Rotblat, J. Steinberger and B. Udgaonkar (Eds.), *A Nuclear-Weapon-Free World: Desirable? Feasible?*, Westview Press, (1993).

(75) The United Methodist Council of Bishops, *In Defense of Creation: The Nuclear Crisis and a Just Peace*, Graded Press, Nashville, (1986).
(76) S.R. Weart, *Nuclear Fear: A History of Images*, Harvard University Press, (1988).
(77) C. Langley, *Soldiers in the Laboratory: Military Involvement in Science and Technology and Some Alternatives*, Scientists for Global Responsibility, (2005).
(78) M.T. Klare, *Blood and Oil: The Dangers and Consequences of America's Growing Dependency on Imported Petroleum*, Metropolitan Books, New York, (2004); paperback, Owl Books, (2005).
(79) M.T. Klare, *Resource Wars: The New Landscape of Global Conflict*, reprint edition, Owl Books, New York, (2002).
(80) M. Renner, *The Anatomy of Resource Wars*, Worldwatch Paper #162, Worldwatch Institute, (2002).
(81) W.B. Gallie, *Understanding War: Points of Conflict*, Routledge, London, (1991).
(82) R. Falk and S.S. Kim, eds., *The War System: An Interdisciplinary Approach*, Westview, Boulder, CO, (1980).
(83) J.D. Clarkson and T.C. Cochran, eds., *War as a Social Institution*, Colombia University Press, New York, (1941).
(84) S. Melman, *The Permanent War Economy*, Simon and Schuster, (1974).

Chapter 12

SCIENCE: ITS OPPORTUNITIES AND DANGERS

12.1 Science as organized knowledge

The predictive value of statements

Science is defined as organized knowledge. Over the centuries, humans have worked to build up a body of verifiable knowledge that is as free as possible from error, although of course it is not at all complete. By "verifiable", one means that the knowledge can be used to correctly predict future events. The predictive value of a statement is a measure of its degree of truth[1]. For example, Newton's laws of motion can be used to predict the future motions of the planets; but Newton's laws are only approximately true. Einstein's relativistic laws of gravitation and motion are more exact than Newton's, as can be verified by observing the finer details of the motion of the planet Mercury; but even Einstein's laws may need improvement in the future. In this way, the body of verifiable knowledge slowly expands. All scientists recognize that our knowledge is far from complete. Newton's comment was, "I know not how I may appear to the world, but to myself I seem to have been only like a boy playing on the seashore and diverting myself in now and then finding a smoother pebble or a prettier shell than ordinary, whilst the great ocean of truth lay all undiscovered before me."

The body of knowledge found in this way has turned out to have a high degree of unity and harmony. Historically, science has been divided into separate disciplines - mathematics, astronomy, physics, chemistry, biology, and so on. In more recent times, the close relationships between these fields has led to hybrid disciplines such as biochemistry, biophysics, biophysical

[1]Suppose that you have a small box which you claim contains a coin. Various predictions follow from this statement, the prediction that if you open the box, you will see the coin, or the prediction that if you shake the box, it will rattle, and so on. Thus the statement is verifiable. When a sufficient number of tests have been made, the statement can be incorporated into the body of verified knowledge.

chemistry, astrophysics, physical chemistry, chemical physics, mathematical biology, etc. In the social sciences too, interdisciplinary fields have become increasingly important. For example, computerized systems analysis is used in economic modeling and radioisotope methods, such as radiocarbon dating, are enormously helpful in archeology.

The fundamental laws of nature (discovered until now) seem to have great simplicity and beauty. The famous English theoretical physicist P.A.M. Dirac once wrote a paper in the Canadian Journal of Physics in which he pointed out that the fundamental laws of physics have all turned out to have great mathematical beauty. This being so, Dirac maintained, we ought to follow our sense of mathematical beauty when we search for new physical laws. Incidentally this method worked for Dirac! He made many pioneering discoveries by following his highly-developed sense of mathematical beauty.

Reductionism versus Holism

The fact that interdisciplinary studies are increasingly important in science is an indicator of the great degree of harmony and unity that we can observe in our rapidly-growing body of verifiable knowledge. When chemical systems are examined deeply, we find that their behavior is determined by the laws of physics. For example, all the details of chemical bonding can be explained by quantum theory, a branch of theoretical physics. Biochemistry is just the chemistry of biologically important molecules, and the facts of physiology can be explained in terms of biochemistry. Thus in a way, physiology can be said to be reducible to physics. This way of looking at complex systems is called "reductionism", and although it has a certain degree of truth, it is also misleading.

It is true that in principle one could write down a quantum mechanical equation whose solution would tell us about he behavior of a living organism; but in practice, such an approach would be folly. The equation would be far too long and complex even to write down, and of course impossible to solve. Thus in principle reductionism gives us a link between different disciplines, but in practice, each field has its own methods and its own abstractions.

The models used by various branches of science are abstractions, i.e. the models are simpler than the real systems that they represent, just as a model of a ship is simpler than the ship itself. Scientific models focus on particular aspects of reality, and the aspects on which it is useful to focus differ from one field to another. Holistic models are often needed describe the behavior of complex systems

Despite its limitations, the reductionist viewpoint does give us a second way of testing a statement before it is incorporated into the great body of organized knowledge. Does the statement harmonize with what we already know? If not, it should be viewed with suspicion. We should not be reject such a statement outright, but it should be extremely carefully tested for predictive value, and great efforts should be made to see whether, in some deeper way, it can be made to harmonize with the rest of knowledge.

Induction versus deduction

Charles Darwin tried as much as possible to use inductive (Baconian) methods in his work, and he remarks in his autobiography that "science consists in arranging facts in such a way that general conclusions can be drawn from them". By contrast, mathematics is very largely a deductive science. For example, in Euclid's *Elements*, a few axioms are assumed, and the remainder of the book develops conclusions that follow deductively from the axioms. Both induction and deduction are of value in science. Without induction, the results of deduction from axioms would not necessarily be related to the real world. Without deduction, the wealth of conclusions following from a set of observations might be overlooked.

Is science value-free?

The laws of nature would be the same even if there were no humans in the universe. Science is an attempt to understand these laws, and therefore it is in some degree independent of human values. If there are intelligent beings in other galaxies, they presumably have a version of Newton's laws that is not too different from our own.

On the other hand, as the feminist philosopher Sandra Harding has pointed out, science consists of patches of light, surrounded by large areas of darkness. The patches of light are topics that humans have thought worthwhile to investigate, and their placement reflects human values. This is particularly true of applied science. In fundamental science, the agenda is set by the internal logic of the field, but in applied science, the agenda reflects social needs.

"What is feminist science?", Sandra Harding asked in a recent lecture, "Is it 'electrons with ear-rings'?" No, she explained, it is science dealing with problems that are important to women. Each group of people has a different set of topics that they consider to be important, and hence a different pattern of light and darkness in their body of organized knowledge. For example, it is important for the indigenous peoples of the Amazon rain forests to know which plants are edible, which poisonous and which medicinal; hence their body of organized knowledge (i.e. their science) includes this information.

The social impact of science during the last three centuries has been enormous. It was the driving force behind the Industrial Revolution. It produced the inequalities of the colonial era. Progress in science and technology lowered death rates, and thus produced the population explosion. New weapons based on scientific progress led to the horrors of two world wars. Technological progress has also led to unprecedentedly high standards of living in the developed countries. Because of the enormous social impact of science, it is becoming less and less possible to regard it as value-free. Scientists need to be more conscious of the impact of their work on the general welfare of humankind.

12.2 Exponential growth

The growth of modern science is accelerating because knowledge feeds on itself: A new idea or a new development may lead to several other innovations, which can in turn start an avalanche of change. For example, the quantum theory of atomic structure lead to the invention of transistors, which made the development of high-speed digital computers possible. Computers have not only produced further developments in quantum theory; they have also revolutionized many other fields. X-ray crystallographic analysis of the structure of large biologically important molecules was made practical by computers. This in turn led to the discovery of the structure of DNA and proteins, and the foundation of modern molecular biology.

The growth law which follows from this type of relationship is exponential; and in fact, the number of scientific articles published per year increased exponentially during the 20th century, doubling every fifteen years or so. The exponential growth of technology was the driving force behind the other exponentially increasing graphs which can be made, such as the graphs of population growth and the growth of international trade.

Since the number of scientific articles published doubled each decade and a half for many years, the number of scientists must also have had roughly the same doubling time. Obviously, exponential growth of any kind cannot continue indefinitely, and in fact, the growth of science and engineering is now slowing down, if one uses as a measure the number of students completing their education in these fields.

The long period of exponential growth of science produced some changes in its character: Prior to the second half of the 20th century, science was very largely conducted in academic environments. Expenses were limited, and could usually be covered by university endowments. Some scientists, for example Napier, Boyle, Lavoisier, Cavendish and Darwin, were independently wealthy. They pursued their studies because they found them intellectually exciting. In general, the aim of science was not to make money but to discover the laws of nature. Scientific knowledge, once discovered, was always published, and was freely available to everyone throughout the world. Scientists felt themselves to be amply rewarded by the esteem of their peers.

However, during the second half of the 20th century, the enormous growth of science produced some changes in this pattern. Science and technology had shown themselves to be of great practical value to human society, and society came to expect practical results. Furthermore, expenses were increasing drastically, partly because of the rapidly growing number of scientists, and partly because much more expensive and complex equipment was being used.

Today, a typical scientist spends much of his or her time applying for grants to support the research of graduate students. Such applications emphasize the practical results that are likely to flow from the projects. Much research is conducted by industrial laboratories, and the results are not freely published. Patents are considered to be necessary to make research worthwhile. Science is expected to pay for itself by yielding practical results.

12.3 Patrons of science and engineering

Soldiers in the laboratory

Ideally science ought to be the disinterested study of nature, combined with a search for practical applications of benefit to humanity as a whole. Ideally, universities ought to be the eyes, mind and conscience of society. However, during the last few decades, these ideals have been progressively eroded. Dr. Chris Langley's important book, *Soldiers in the laboratory*[2] describes the destruction of academic ideals by the pervasive influence of the military industrial complex.

Although the details of Langley's report deal mainly with the United Kingdom, the book also touches on what is happening to an even larger extent in the United States, and also in other countries. Academic scientists and engineers will have noticed that in recent years, tenured positions have become increasingly scarce; increasingly they are expected to finance research and salaries by means of contracts with industry. Research nar-

[2](published by Scientists for Global Responsibility, (2005), and available at www.sgr.org)

Fig. 12.1 *The Chinese ideogram for "crisis" has two parts, one meaning "danger", and the other, "opportunity".*

rowly aimed at wealth-production is encouraged rather than fundamental science.

So far so bad, but this is only the beginning of the story. As Langley amply documents, the industrial contracts that finance so much of today's research are often backed by military money and aimed at military goals. In Langley's Chapter 4 he looks in detail at four cases where science and engineering are being cynically used in the service of war:

(1) Biological sciences and the military
(2) Nanotechnology - from nanotubes to the battlefield
(3) Missile Defense and the securing of dominance
(4) A new generation of nuclear weapons

These goals, Langley shows, are not only destructive in themselves, but resources used on arms are diverted from research on the urgent problems facing humanity - food security, clean water, the resurgence of infectious disease, poverty, climate change and so on. Langley pleads for a wider definition of security, where solutions to the urgent problems just named would be seen as contributing to global stability and safety.

UNESCO as a patron of science and technology

If science and technology have grown so large that they can no longer be financed by university endowments, and if the military-industrial complex is a highly undesirable patron, who will pay the bills? Industries also have defects as patrons: They aim at new products that are not necessarily needed, and they neglect much-needed areas of research if they are not profitable. For example, pharmaceutical companies neglect research on tropical diseases because this field not seen as being profitable. Furthermore, results found by industrial research laboratories are often not published and not freely available throughout the world; often they are protected by patents.

There are several reasons why the United Nations Educational Scientific and Cultural Organization (UNESCO) would be an ideal patron for science and engineering if it only had enough money to fill this role. Not being motivated by the desire for profit, UNESCO could direct more funds to urgently needed research on the most pressing problems that face humanity[3]. Secondly, not being profit-oriented, UNESCO could sponsor fundamental research. Thirdly, since UNESCO is an international organization, it could encourage transnational cooperation, one of the best methods of peace-building. Finally, the prestige gained from the sponsorship of science and technology would help the United Nations in situations where it is called on to solve difficult political problems.

No real government lacks the power of taxation. We cannot expect the United Nations to solve global problems effectively without giving it this power. Some of the tax possibilities discussed in Chapter 11, especially the Tobin Tax, could give the United Nations, and hence UNESCO, enough money to become the most important future patron of science and engineering.

12.4 Rapidly growing fields

The history of 20th century science and science-based technology has been one of remarkable and unexpected discoveries and inventions, occurring at an ever-increasing rate. This being so, what principles can guide us if we are so rash as to try to predict the trends which technology will follow during the 21st century?

Lacking other clues, we can, of course, predict that those disciplines

[3]Tom Børsen Hansen of the University of Copenhagen has given the name "Grassroots Science" to research aimed at solving humanity's most urgent problems.

which are at present developing most dramatically will continue to do so. At present, the two most rapidly-developing fields are biotechnology and information technology (microelectronics). These two fields will undoubtedly continue to develop with astonishing speed; and one can already foresee some areas of interaction and overlap between them.

Secondly, we can predict that a severe crisis will occur during the 21st century, as a global population of completely unprecedented size exhausts the earth's limited supply of non-renewable resources. This crisis will present a challenge to the technology of the new century: New sources of food will have to be found; new materials will have to be developed to replace scarce minerals; and renewable energy sources will have to be introduced on a large scale to replace fossil fuels. Technology and applied science in the future will be shaped by these pressing needs.

Biotechnology

During the period since World War II, our understanding and control of fundamental biological processes has grown dramatically. Techniques from the physical sciences (such as X-ray crystallography, electron microscopy and nuclear magnetic resonance) gave molecular biologists a detailed picture of the way in which DNA controls the synthesis of proteins, as well as the role of proteins in controlling the synthesis and metabolism of smaller molecules. The genetic code was broken, and automatic techniques were developed for determining the base sequences of DNA and RNA molecules and the amino acid sequences of proteins. Molecular force-field computer programs were also developed which were able to calculate the tertiary conformations of proteins from the amino acid sequences.

In the 1970's, recombinant DNA techniques made it possible for researchers to mix the genetic repertoires of different species. Thus, for the first time, humans achieved direct control over the process of evolution. The genes of mice and men could be spliced together into new, man-made forms of life. With the help of restriction enzymes and DNA ligase, the gene for the synthesis of a particular protein could be spliced into a plasmid and inserted into a bacterium, whose progeny could be cultured, thus producing the chosen protein cheaply and in large quantities. In the 1980's, a number of genetic-engineering products reached the market. These included hormones, vaccines, amino acids, antibiotics, pesticides, pesticide-resistant plants, cloned livestock, improved yeasts, cellulose-digesting bacteria, and a nitrogen-fixation enzyme. It seems likely that continued work in this direction will soon produce gene-spliced nitrogen-fixing bacteria which are able to live on the roots of wheat and rice plants, thus making nitrogen-containing fertilizers unnecessary.

In 1976, the first commercial genetic engineering company (Genentech) was founded. In 1980, the initial public offering of Genentech stock set a Wall Street record for the fastest increase of price per share. In 1981, another genetic engineering company (Cetus) set a Wall Street record for the largest amount of money raised in an initial public offering (125 million U.S. dollars). During the same years, Japan's Ministry of International Trade and Technology declared 1981 to be "The Year of Biotechnology"; and England, France and Germany all targeted biotechnology as an area for special development.

In 1983, the polymerase chain reaction was introduced, a technique which made it possible to reproduce a given segment of DNA in large quantities and with little effort. The technique also made it possible to determine whether or not a given oligonucleotide sequence is present in a sample of DNA. An ambitious program was initiated to map the entire human genome. It is anticipated that with the help of a placental biopsy, the human genome map, and the polymerase chain reaction, it will be possible to say a great deal about the unborn child of a woman who is only 10 weeks pregnant. It will thus be possible to diagnose and treat genetic diseases at a very early stage through DNA technology.

The possibility of extensive genetic screening raises ethical problems which require both knowledge and thought on the part of the public. An expectant mother, in an early stage of pregnancy, often has an abortion if the fetus is found to carry a serious genetic defect. But with more knowledge, many more defects will be found. Where should the line be drawn between a serious defect and a minor one?

The cloning of genes for lethal toxins also needs serious thought and public discussion. From 1976 to 1982, this activity was prohibited in the United States under the NIH Guidelines. However, in April, 1982, the restriction was lifted, and by 1983, the toxins being cloned included several aflatoxins, lecithinase, cytochalasins, ochratoxins, sporidesmin, T-2 toxin, ricin and tremogen. Although international conventions exist under which chemical and biological weapons are prohibited, there is a danger that nations will be driven to produce and stockpile such weapons because of fear of what other nations might do.

Finally, the release of new, transgenic species into the environment requires thought and caution. Much benefit can come, for example, from the use of gene-spliced bacteria for nitrogen fixation or for cleaning up oil spills. However, once a gene-spliced microorganism has been released, it is virtually impossible to eradicate it; and thus the change produced by the release of a new organism is permanent. Permanent changes in the environment should not be made on the basis of short-term commercial considerations, nor indeed on the basis of short-term considerations of any kind; nor should such decisions be made unilaterally by single nations, since new organisms can easily cross political boundaries.

The recently completed Human Genome Project is expected to make possible prenatal diagnosis of many inherited diseases. For example, the gene for cystic fibrosis has been found; and DNA technology makes it possible to detect the disease prenatally.

In agriculture, we can anticipate that genetic engineering will be able to produce plant foods nutritionally designed to make meat unnecessary. In agriculture as in medicine, ethical questions will arise, as our control of the evolutionary process increases: How accurately can we predict the ecological effect of manmade species when we release them into the global environment? Is the only purpose of nonhuman species to serve human ends? To what extent is it allowable to make nature subservient to human goals?

The rapid development of biotechnology has given humans enormous power over the fundamental mechanisms of life and evolution. But is society mature enough to use this power wisely and compassionately?

The Asilomar Conference

In February, 1975, more than 100 leading molecular biologists from many parts of the world met at the Asilomar Conference Center near Monterey, California, to discuss safety guidelines for recombinant DNA research. There was an almost unanimous consensus at the meeting that, until more was known about the dangers, experiments involving cloning of DNA should make use of organisms and vectors incapable of living outside a laboratory environment.

The Asilomar Conference also recommended that a number of experiments be deferred. These included cloning of recombinant DNA derived from highly pathogenic organisms, or containing toxin genes, as well as large-scale experiments using recombinant DNA able to make products potentially harmful to man, animals or plants.

The Asilomar recommendations were communicated to a special committee appointed by the U.S. National Institutes of Health; and the committee drew up a set of guidelines for recombinant DNA research. The NIH Guidelines went into effect in 1976; and they remained in force until 1979. They were stricter than the Asilomar recommendations regarding cloning of DNA from cancer-producing viruses; and this was effectively forbidden by the NIH until 1979. The problem, of course, is that the NIH Guidelines were effective only for research conducted within the United States and funded by the U.S. government. Biotechnology rapidly grew beyond these boundaries, and hence beyond control by the guidelines.

Information technology and microelectronics

During the last three decades, the cost of computing has decreased exponentially by more than 20 percent per year; and the computer industry has grown at almost the same exponential rate. This enormous increase in the speed and economy of computers has been due to the introduction of transistors and to the micro-miniaturization of integrated circuits.

The limiting factor in computer speed has been found to be the time needed for an electrical signal to propagate from one part of the central processing unit to another. Since electrical impulses propagate with the speed of light, this time is extremely small; but nevertheless, it is the limiting factor in the speed of electronic computers. In order to reduce the propagation time, computer designers tried to make the central processing units very small; and the result was the development of integrated circuits and microelectronics.

Integrated circuits were developed, in which single circuit elements were not manufactured separately, but instead the whole circuit was made at one time. An integrated circuit is a multilayer sandwich-like structure, with conducting, resisting and insulating layers interspersed with layers of germanium or silicon, "doped " with appropriate impurities. At the start of the manufacturing process, an engineer makes a large drawing of each layer. For example, the drawing of a conducting layer would contain pathways which fill the role played by wires in a conventional circuit, while the remainder of the layer would consist of areas destined to be etched away by acid.

The next step is to reduce the size of the drawing and to multiply it photographically. The pattern of the layer is thus repeated many times, like the design on a piece of wallpaper. The multiplied and reduced drawing is then focused through a reversed microscope onto the surface to be etched.

Successive layers are built up by evaporating or depositing thin films of the appropriate substances onto the surface of a silicon or germanium wafer. If the layer being made is to be conducting, the surface might consist of an extremely thin layer of copper, covered with a photosensitive layer called a "photoresist". On those portions of the surface receiving light from the pattern, the photoresist becomes insoluble, while on those areas not receiving light, the photoresist can be washed away.

The surface is then etched with acid, which removes the copper from those areas not protected by photoresist. Each successive layer of a wafer is made in this way, and finally the wafer is cut into tiny "chips", each of which corresponds to one unit of the wallpaper-like pattern. Although the area of a chip may be much smaller than a square centimeter, the chip can contain an extremely complex circuit.

Fig. 12.2 *An Internet café.*

A typical programmable minicomputer or "microprocessor", manufactured in the 1970's, could have 30,000 circuit elements, all of which were contained on a single chip. By 1989, more than a million transistors were being placed on a single chip; and by 2000, the number reached 42,000,000. As a result of miniaturization and parallelization, the speed of computers has risen exponentially.

Computer disk storage has also undergone a remarkable development. In 1987, the magnetic disk storage being produced could store 20 million bits of information per square inch; and even higher densities could be achieved by optical storage devices. The storage available for a constant price is currently doubling every 18 months.

In 1961, an MIT graduate student named Leonard Kleinrock submitted a proposal for Ph.D. thesis entitled "Information Flow in Large Communication Nets". Kleinrock's proposal marked the start of a line of development that ultimately lead to the Internet. In his thesis proposal he wrote: "The nets under consideration consist of nodes, connected to each other by links. The nodes receive, sort, store, and transmit messages that enter and leave via the links. The links consist of one-way channels, with fixed capacities.

Among the typical systems which fit this description are the Post Office System, telegraph systems, and satellite communication systems."

In telephone communication, there is a direct link between the two speakers. By contrast, in a postal system, there is no such connection - only the addresses of the sender and receiver on the package of information. In the Internet, and in other computer networks based on the package switching principle, a package of information makes its way from node to node until it reaches its destination.

The enormous success of the package switching principle in computer networks can be measured by the fact that between 1990 and 1994, the traffic on the Internet doubled every year. Between 1994 and 1996, it leaped from 16.3 trillion bytes per month to 1.5 quadrillion, and it is still growing explosively, revolutionizing the worldwide information-sharing process.

Autoassembly of supramolecular structures

Interestingly, information technology and biology seem to be merging, each field gaining inspiration and help from the other. For example, computer scientists are producing computer hardware and software (neural networks) that mimic the mechanism of the brain; and conversely, neurophysiologists are aided by insights from the field of artificial intelligence. Another merging of biology and information technology can be found in the new field of bioinformatics, where computer techniques are use to analyze the rapidly-growing libraries of information about DNA, RNA and protein sequences.

Miniaturization is the key to increasing computer speeds, but miniaturization using conventional methods has limits, and these are now being approached. Therefore designers of integrated circuits are turning for inspiration to the principles of molecular complementarity and autoassembly, which have long been familiar in biology.

Biologists have long been aware that the language of molecular complementarity is the chief mechanism by which information is stored and transfered in biological systems. Biological molecules have physical shapes and patterns of excess charge, as well as patterns of polarizable groups and reactive groups, which are recognized by complementary molecules because they fit together, in much the same way as a key fits the shape of a lock. We can see an example of biological "lock and key" fitting in the recognition of an antigen by its specific antibody. Other examples include enzyme-substrate specificity, the specificity of base pairs in DNA and RNA, and the autoassembly of structures such as viruses and subcellular organelles.

The tobacco mosaic virus, for example, can be decomposed into its constituent molecules, and the protein and RNA can be separated and put into separate bottles. We can put the bottles, one containing RNA, the other protein molecules, onto the laboratory shelf and leave them there for months or years. At a later time, we can take the bottles down again and shake the protein and RNA molecules together in water. If we do so, they will spontaneously assemble themselves into new infective tobacco mosaic virus particles! The process of autoassembly is analogous to crystallization, except that the structure formed is more complex than an ordinary crystal.

Self-organization by molecular complementarity and pattern recognition is one of the most important universal mechanisms of biology. Today, chemists, nonotechnologists and integrated circuit designers would like to utilize the principle of autoassembly to design and produce structures that have the approximate size of virus particles, i.e., nanostructures.

Artificial life

The question of whether we are now able to produce artificial life depends on how the term is defined. In a sense, one might maintain that all domestic species can be called "artificial life", since humans have played such an important role in their evolution. However, the chimeras created by Steen Willadsen come closer to what most people would call "artificial life". Willadsen is famous for having made the first reproducible clone of a mammal in 1984. Later, working in Cambridge, England, he made a series sheep-goat chimeras. In each of these animals, some cells have the genetic characteristics of sheep, while other cells carry goat genes.

In order to make these sheep-goat mosaics, Steen Willadsen operated under a microscope on sheep and goat embryos at the 8-cell stage. At this stage, the embryos are surrounded by a transparent casing called the *zona pelucida.* Willadsen cut open this casing on an embryo and removed all the cells. Then, into the empty *zona pelucida*, he put four cells from a sheep embryo and four from a goat embryo. The chimeral embryo thus produced developed into a healthy adult animal in which some patches of tissue were descended from the four sheep cells while other patches derived from the four goat cells. Willadsen's motive in making the chimeras was to cast light on the the mechanism by which the information on the genome of a growing embryo is translated into morphology - one of the most interesting problems in biology.

It is now commonplace to introduce human genes into farm animals in order to produce medically valuable human proteins, such as insulin, blood-clotting factors, collagen, fibrinogin, fertility hormones and serum albumin. New species of transgenic animals and plants, carrying the genes of two or more species, are quite common today in laboratory environments. In some cases they are also being released into nature. Such transgenic species could justifiably be called "artificial life".

Today computer scientists often evolve software by means of genetic algorithms that make use of Darwinian natural selection. The computer analogue to DNA is a long string of 0's and 1's. In genetic algorithms, these strings are subject to mutations. The mutated strings can "die" or be replicated, depending on whether the mutations are judged to have "survival value".

Methods of artificial evolution inspired by biology have also been applied to computer hardware. Interestingly, circuits designed in this way sometimes work astonishingly well, although no one understands why.

The first international conference on artificial life took place in 1987. It was organized by Christopher Langton of the Center for Nonlinear Studies at Los Alamos, New Mexico, and was called the "Interdisciplinary Workshop on the Synthesis and Simulation of Living Systems". In an announcement for the conference, which appeared in the Scientific American, Langton stated that "Artificial life is the study of artificial systems that exhibit behavior characteristic of natural living systems...The ultimate goal is to extract the logical form of living systems. Microelectronic technology and genetic engineering will soon give us the capability to create new life *in silico* as well as *in vitro.* This capacity will present humanity with the most far-reaching technical, theoretical, and ethical challenges it has ever confronted. The time seems appropriate for a gathering of those involved in attempts to simulate or synthesize aspects of living systems." The conferences have been repeated every year since 1987, and Langton now edits a journal called *Artificial Life.*

12.5 Science and technology out of control?

Science loses its innocence

To the pioneers of science, during the Renaissance, the Enlightenment, the Industrial Revolution and the early 20th century, it seemed that the accumulation of knowledge could only be beneficial to humankind. Condorcet spoke of humanity's "...march towards a future state of absolute perfection", and "...an era when the human race will have attained improvements, of which we can at present scarcely form a conception." Godwin visualized a future utopia of universal plenty where "No man would be the enemy of his neighbor, for they would have nothing to contend; and of consequence philanthropy would assume the empire which reason assigns her. Mind would be delivered from her perpetual anxiety about corporal support, and free to expatiate in the field of thought which is congenial to her. Each man would assist the inquiries of all." Darwin spoke of his "...burning zeal to add even the most humble contribution to the noble structure of Natural Science." The principles of science, once discovered, could never be lost, and (it was believed) they would lead humanity forward to a golden age.

Marie and Pierre Curie's heroic efforts to isolate radium were motivated by the belief that all scientific knowledge must be beneficial. Madame Curie, the first great woman scientist, was portrayed by newspapers as a great humanitarian, since it had been discovered that radium could be used for treating cancer. Indeed, the motives which inspired Marie and Pierre Curie were both humanitarian and idealistic. They did not know that radium is a dangerous element, capable of causing cancer as well as curing it; and they could not foresee that research on radioactivity would eventually lead to nuclear weapons.

With the tragic nuclear destruction of Hiroshima and Nagasaki in 1945, science lost its innocence and revealed itself to be double-edged, capable of doing great harm as well as good. The Nobel laureate biochemist Albert Szent-Györgyi eloquently summarized the situation:

"The story of man consists of two parts, divided by the appearance of modern science at the turn of the century. In the first period, man lived in the world in which his species was born and to which his senses were adapted. In the second, man stepped into a new, cosmic world to which he was a complete stranger... The forces at man's disposal were no longer terrestrial forces, of human dimension, but were cosmic forces, the forces which shaped the universe. The few hundred Fahrenheit degrees of our flimsy terrestrial fires were exchanged for the ten million degrees of the atomic reactions which heat the sun."

"This is but a beginning, with endless possibilities in both directions - a building of a human life of undreamt of wealth and dignity, or a sudden end in utmost misery. Man lives in a new cosmic world for which he was not made. His survival depends on how well and how fast he can adapt himself to it, rebuilding all his ideas, all his social and political institutions."

"...Modern science has abolished time and distance as factors separating nations. On our shrunken globe today, there is room for one group only - the family of man."

Respect for natural evolution

Although it is impossible to foresee in detail what the discoveries of science in the distant future will be, we can at least predict that humans will acquire great power over biological processes, and over the process of evolution, including human evolution. With these great powers will go equally great dangers and responsibilities.

When we design a chair or a table, or for that matter a new transgenic organism, we do it in a way that is totally different from the process of natural evolution. In nature, trial and error has shown what works best, and the trials have been made over a period of more than three billion years! The vast scale of these evolutionary trials, and the vast period of time over which they have conducted deserve our deepest respect. Natural evolution has produced communities of plants and animals that are fine-tuned to live, and to live with each other. Our clumsy attempts at technological design of organisms can produce environmental disasters if we do not proceed with extreme caution. This is not to say that we should reject biotechnology as a means of (for example) solving the problem of global food security. But we should be aware of the dangers.

12.6 The need for ethics

The impossibility of "Star Wars"

As everyone knows, there is a popular series of science fiction films entitled "Star Wars". In this series of films we see societies that are scientifically advanced but ethically primitive. The characters fly among the galaxies in technologically advanced spaceships, but spend most of their time fighting. This is an impossibility, since any scientifically advanced civilization that does not achieve ethical and political maturity to to match its technical progress will quickly destroy itself, as human civilization now is in danger of doing.

The present crisis of civilization is caused by the amazing speed of technical progress, a speed so rapid that not only genetic evolution, but also the non-technical aspects of cultural evolution are left far behind. As Szent-Györgyi said, "Man lives in a new cosmic world for which he was not made. His survival depends on how well and how fast he can adapt himself to it, rebuilding all his ideas, all his social and political institutions."

Global ethics

If human civilization is to survive, it is absolutely essential that we establish a new system of global ethics to supplement our narrow national loyalties. It is interesting to notice that many of the great ethical teachers of the past lived at a time when society was undergoing a transition from a tribal lifestyle to life in larger groups - cities, nations and even cosmopolitan nations. The ethical rules of behavior that early philosophers and religious leaders introduced helped humans to live successfully in large groups. Today we urgently need new ethical principles that will embrace the entire world, building solidarity across all boundaries. Only when a truly global ethical system is in place can we be confident that science will be beneficial rather than harmful.

A non-anthropocentric component for ethics

The ethics of the future will need non-anthropocentric elements. We must regard other species than our own as worthy of preservation not just because they are useful or potentially useful to humans, but for their own sakes. Reverence for all life must be central to the ethics of the future.

Inanimate nature also deserves respect; for example, topsoil needs to be preserved from erosion; groundwater, lakes, rivers, seas and oceans must be protected from pollution, the atmosphere from greenhouse gasses, and the ozone layer from hydroflurocarbons. Since the future will be a time of energy scarcity and resource scarcity, humans will need to derive more of their pleasures from the appreciation of unspoiled nature and less from the accumulation of manufactured goods. Therefore one of the ethical principles of the future should be the duty to protect sites of natural beauty from degradation. Members of our species will need to regard themselves not as inheritors of the earth, or owners of the earth, but as a modest part of the natural environment.

Finally, the rights of future generations should hold a high place in the ethics of the future.

Suggestions for further reading

(1) T. Børsen Hansen, *Grassroots science - an ISYP Ideal?*, ISYP Journal on Science and World Affairs, **1**, 61-72, (2005).
(2) S. Harding, *Whose Science? Whose Knowledge?: Thinking From Women's Lives*, (1991).
(3) H. Lodish, A. Berk, S.L. Zipursky, P. Matsudaira, D. Baltimore, and J. Darnell, *Molecular Cell Biology*, 4th Edition, W.H. Freeman, New York, (2000).
(4) Lily Kay, *Who Wrote the Book of Life? A History of the Genetic Code*, Stanford University Press, Stanford CA, (2000).
(5) Sahotra Sarkar (editor), *The Philosophy and History of Molecular Biology*, Kluwer Academic Publishers, Boston, (1996).
(6) James D. Watson et al. *Molecular Biology of the Gene*, 4th Edition, Benjamin-Cummings, (1988).
(7) J.S. Fruton, *Proteins, Enzymes, and Genes*, Yale University Press, New Haven, (1999).
(8) S.E. Lauria, *Life, the Unfinished Experiment*, Charles Scribner's Sons, New York (1973).

(9) A. Lwoff, *Biological Order*, MIT Press, Cambridge MA, (1962).
(10) James D. Watson, *The Double Helix*, Athenium, New York (1968).
(11) F. Crick, *The genetic code*, Scientific American, **202**, 66-74 (1962).
(12) F. Crick, *Central dogma of molecular biology*, Nature, **227**, 561-563 (1970).
(13) David Freifelder (editor), *Recombinant DNA*, Readings from the Scientific American, W.H. Freeman and Co. (1978).
(14) James D. Watson, John Tooze and David T. Kurtz, *Recombinant DNA, A Short Course*, W.H. Freeman, New York (1983).
(15) Richard Hutton, *Biorevolution, DNA and the Ethics of Man-Made Life*, The New American Library, New York (1968).
(16) Martin Ebon, *The Cloning of Man*, The New American Library, New York (1978).
(17) Sheldon Krimsky, *Genetic Alchemy: The Social History of the Recombinant DNA Controversy*, MIT Press, Cambridge Mass (1983).
(18) M. Lappé, *Germs That Won't Die*, Anchor/Doubleday, Garden City N.Y. (1982).
(19) M. Lappé, *Broken Code*, Sierra Club Books, San Francisco (1984).
(20) President's Commission for the Study of Ethical Problems in Medicine and Biomedical and Behavioral Research, *Splicing Life: The Social and Ethical Issues of Genetic Engineering with Human Beings*, U.S. Government Printing Office, Washington D.C. (1982).
(21) U.S. Congress, Office of Technology Assessment, *Impacts of Applied Genetics - Microorganisms, Plants and Animals*, U.S. Government Printing Office, Washington D.C. (1981).
(22) W.T. Reich (editor), *Encyclopedia of Bioethics*, The Free Press, New York (1978).
(23) Martin Brown (editor), *The Social Responsibility of the Scientist*, The Free Press, New York (1970).
(24) B. Zimmerman, *Biofuture*, Plenum Press, New York (1984).
(25) John Lear, *Recombinant DNA, The Untold Story*, Crown, New York (1978).
(26) B. Alberts, D. Bray, J. Lewis, M. Raff, K. Roberts and J.D. Watson, *Molecular Biology of the Cell*, Garland, New York (1983).
(27) C. Woese, *The Genetic Code; The Molecular Basis for Genetic Expression*, Harper & Row, New York, (1967).
(28) F.H.C. Crick, *The Origin of the Genetic Code*, J. Mol. Biol. **38**, 367-379 (1968).
(29) M.W. Niernberg, *The genetic code: II*, Scientific American, **208**, 80-94 (1962).
(30) L.E. Orgel, *Evolution of the Genetic Apparatus*, J. Mol. Biol. **38**, 381-393 (1968).

(31) Melvin Calvin, *Chemical Evolution Towards the Origin of Life, on Earth and Elsewhere*, Oxford University Press (1969).
(32) R. Shapiro, *Origins: A Skeptic's Guide to the Origin of Life*, Summit Books, New York, (1986).
(33) J. William Schopf, *Earth's earliest biosphere: its origin and evolution*, Princeton University Press, Princeton, N.J., (1983).
(34) J. William Schopf (editor), *Major Events in the History of Life*, Jones and Bartlet, Boston, (1992).
(35) Robert Rosen, *Life itself: a comprehensive inquiry into the nature, origin and fabrication of life*, Colombia University Press, (1991).
(36) R.F. Gesteland, T.R Cech, and J.F. Atkins (editors), *The RNA World*, 2nd Edition, Cold Spring Harbor Laboratory Press, Cold Spring Harbor, New York, (1999).
(37) C. de Duve, *Blueprint of a Cell*, Niel Patterson Publishers, Burlington N.C., (1991).
(38) C. de Duve, *Vital Dust; Life as a Cosmic Imperative*, Basic Books, New York, (1995).
(39) F. Dyson, *Origins of Life*, Cambridge University Press, (1985).
(40) S.A. Kaufman, *Antichaos and adaption*, Scientific American, **265**, 78-84, (1991).
(41) S.A. Kauffman, *The Origins of Order*, Oxford University Press, (1993).
(42) F.J. Varela and J.-P. Dupuy, *Understanding Origins: Contemporary Views on the Origin of Life, Mind and Society*, Kluwer, Dordrecht, (1992).
(43) Stefan Bengtson (editor) *Early Life on Earth; Nobel Symposium No. 84*, Colombia University Press, New York, (1994).
(44) Herrick Baltscheffsky, *Origin and Evolution of Biological Energy Conversion*, VCH Publishers, New York, (1996).
(45) J. Chilea-Flores, T. Owen and F. Raulin (editors), *First Steps in the Origin of Life in the Universe*, Kluwer, Dordrecht, (2001).
(46) R.E. Dickerson, Nature **283**, 210-212 (1980).
(47) R.E. Dickerson, Scientific American **242**, 136-153 (1980).
(48) C.R. Woese, *Archaebacteria*, Scientific American **244**, 98-122 (1981).
(49) N. Iwabe, K. Kuma, M. Hasegawa, S. Osawa and T. Miyata, *Evolutionary relationships of archaebacteria, eubacteria, and eukaryotes inferred phylogenetic trees of duplicated genes*, Proc. Nat. Acad. Sci. USA **86**, 9355-9359 (1989).
(50) C.R. Woese, O. Kundler, and M.L. Wheelis, *Towards a Natural System of Organisms: Proposal for the Domains Archaea, Bacteria and Eucaria*, Proc. Nat. Acad. Sci. USA **87**, 4576-4579 (1990).
(51) W. Ford Doolittle, *Phylogenetic Classification and the Universal Tree*, Science, **284** (1999).

(52) G. Wächterhäuser, *Pyrite formation, the first energy source for life: A hypothesis*, Systematic and Applied Microbiology **10**, 207-210, (1988).
(53) G. Wächterhäuser, *Before enzymes and templates: Theory of surface metabolism*, Microbiological Reviews, **52**, 452-484 (1988).
(54) G. Wächterhäuser, *Evolution of the first metabolic cycles*, Proc. Nat. Acad. Sci. USA **87**, 200-204 (1990).
(55) G. Wächterhäuser, *Groundworks for an evolutionary biochemistry - the iron-sulfur world*, Progress in Biophysics and Molecular Biology **58**, 85-210 (1992).
(56) L.H. Caporale (editor), *Molecular Strategies in Biological Evolution*, Ann. N.Y. Acad. Sci., May 18, (1999).
(57) Werner Arber, *Elements in Microbal Evolution*, J. Mol. Evol. **33**, 4 (1991).
(58) Michael Gray, *The Bacterial Ancestry of Plastids and Mitochondria*, BioScience, **33**, 693-699 (1983).
(59) Michael Grey, *The Endosymbiont Hypothesis Revisited*, International Review of Cytology, **141**, 233-257 (1992).
(60) Lynn Margulis and Dorian Sagan, *Microcosmos: Four Billion Years of Evolution from Our Microbal Ancestors*, Allan and Unwin, London, (1987).
(61) Lynn Margulis and Rene Fester, eds., *Symbiosis as as Source of Evolutionary Innovation: Speciation and Morphogenesis*, MIT Press, (1991).
(62) Charles Mann, *Lynn Margulis: Science's Unruly Earth Mother*, Science, **252**, 19 April, (1991).
(63) Jan Sapp, *Evolution by Association; A History of Symbiosis*, Oxford University Press, (1994)
(64) J.A. Shapiro, *Natural genetic engineering in evolution*, Genetics, **86**, 99-111 (1992).
(65) E.M. De Robertis et al., *Homeobox genes and the vertebrate body plan*, Scientific American, July, (1990).
(66) H. Babbage, *Babbages Calculating Engines: A Collection of Papers by Henry Prevost Babbage*, MIT Press, (1984).
(67) A.M. Turing, *The Enigma of Intelligence*, Burnett, London (1983).
(68) R. Penrose, *The Emperor's New Mind: Concerning Computers, Minds, and the Laws of Physics*, Oxford University Press, (1989).
(69) S. Wolfram, *A New Kind of Science*, Wolfram Media, Champaign IL, (2002).
(70) A.M. Turing, *On computable numbers, with an application to the Entscheidungsproblem*, Proc. Lond. Math. Soc. Ser 2, **42**, (1937). Reprinted in M. David Ed., *The Undecidable*, Raven Press, Hewlett N.Y., (1965).

(71) N. Metropolis, J. Howlett, and Gian-Carlo Rota (editors), *A History of Computing in the Twentieth Century*, Academic Press (1980).
(72) J. Shurkin, *Engines of the Mind: A History of Computers*, W.W. Norten, (1984).
(73) J. Palfreman and D. Swade, *The Drream Machine: Exploring the Computer Age*, BBC Press (UK), (1991).
(74) T.J. Watson, Jr. and Peter Petre, *Father, Son, and Co.*, Bantham Books, New York, (1991).
(75) A. Hodges, *Alan Turing: The Enegma*, Simon and Schuster, (1983).
(76) H.H. Goldstein, *The Computer from Pascal to Von Neumann*, Princeton University Press, (1972).
(77) C.J. Bashe, L.R. Johnson, J.H. Palmer, and E.W. Pugh, *IBM's Early Computers*, (Vol. 3 in the History of Computing Series), MIT Press, (1986).
(78) K.D. Fishman, *The Computer Establishment*, McGraw-Hill, (1982).
(79) S. Levy, *Hackers*, Doubleday, (1984).
(80) S. Franklin, *Artificial Minds*, MIT Press, (1997).
(81) P. Freiberger and M. Swaine, *Fire in the Valley: The Making of the Personal Computer*, Osborne/McGraw-Hill, (1984).
(82) R.X. Cringely, *Accidental Empires*, Addison-Wesley, (1992).
(83) R. Randell editor, *The Origins of Digital Computers, Selected Papers*, Springer-Verlag, New York (1973).
(84) H. Lukoff, *From Dits to Bits*, Robotics Press, (1979).
(85) D.E. Lundstrom, *A Few Good Men from Univac*, MIT Press, (1987).
(86) D. Rutland, *Why Computers Are Computers (The SWAC and the PC)*, Wren Publishers, (1995).
(87) P.E. Ceruzzi, *Reckoners: The Prehistory of the Digital Computer, from Relays to the Stored Program Concept, 1935-1945*, Greenwood Press, Westport, (1983)
(88) S.G. Nash, *A History of Scientific Computing*, Adison-Wesley, Reading Mass., (1990).
(89) P.E. Ceruzzi, *Crossing the divide: Architectural issues and the emergence of stored programme computers, 1935-1953*, IEEE Annals of the History of Computing, **19**, 5-12, January-March (1997).
(90) P.E. Ceruzzi, *A History of Modern Computing*, MIT Press, Cambridge MA, (1998).
(91) K. Zuse, *Some remarks on the history of computing in Germany*, in *A History of Computing in the 20th Century*, N. Metropolis et al. editors, 611-627, Academic Press, New York, (1980).
(92) A.R. Mackintosh, *The First Electronic Computer*, Physics Today, March, (1987).
(93) S.H. Hollingdale and G.C. Tootil,*Electronic Computers*, Penguin Books Ltd. (1970).

(94) A. Hodges, *Alan Turing: The Enegma*, Simon and Schuster, New York, (1983).
(95) Alan Turing, *On computable numbers with reference to the Entscheidungsproblem*, Journal of the London Mathematical Society, **II, 2. 42**, 230-265 (1937).
(96) J. von Neumann, *The Computer and the Brain*, Yale University Press, (1958).
(97) I.E. Sutherland, *Microelectronics and computer science*, Scientific American, 210-228, September (1977).
(98) W. Aspray, *John von Neumann and the Origins of Modern Computing*, M.I.T. Press, Cambridge MA, (1990, 2nd ed. 1992).
(99) W. Aspray, *The history of computing within the history of information technology*, History and Technology, **11**, 7-19 (1994).
(100) G.F. Luger, *Computation and Intelligence: Collected Readings*, MIT Press, (1995).
(101) Z.W. Pylyshyn, *Computation and Cognition: Towards a Foundation for Cognative Science*, MIT Press, (1986).
(102) D.E. Shasha and C. Lazere, *Out of Their Minds: The Creators of Computer Science*, Copernicus, New York, (1995).
(103) W. Aspray, *An annotated bibliography of secondary sources on the history of software*, Annals of the History of Computing **9**, 291-243 (1988).
(104) R. Kurzweil, *The Age of Intelligent Machines*, MIT Press, (1992).
(105) S.L. Garfinkel and H. Abelson, eds., *Architects of the Information Society: Thirty-Five Years of the Laboratory for Computer Sciences at MIT*, MIT Press, (1999).
(106) J. Haugeland, *Artificial Intelligence: The Very Idea*, MIT Press, (1989).
(107) M.A. Boden, em Artificial Intelligence in Psychology: Interdisciplinary Essays, MIT Press, (1989).
(108) J.W. Cortada, *A Bibliographic Guide to the History of Computer Applications, 1950-1990*, Greenwood Press, Westport Conn., (1996).
(109) M. Campbell-Kelly and W. Aspry, *Computer: A History of the Information Machine*, Basic Books, New York, (1996).
(110) B.I. Blum and K. Duncan, editors, *A History of Medical Informatics*, ACM Press, New York, (1990).
(111) J.-C. Guedon, *La Planète Cyber, Internet et Cyberspace*, Gallimard, (1996).
(112) J. Segal, *Théorie de l'information: sciences, techniques et société, de la seconde guerre mondaile à l'aube du XXIe siècle*, Thèse de Doctorat, Université Lumière Lyon II, (1998), (http://www.mpiwg-berlin.mpg.de/staff/segal/thesis/)

(113) S. Augarten, *Bit by Bit: An Illustrated History of Computers*, Unwin, London, (1985).
(114) N. Wiener, *Cybernetics; or Control and Communication in the Animal and the Machine*, The Technology Press, John Wiley & Sons, New York, (1948).
(115) W.R. Ashby, *An Introduction to Cybernetics*, Chapman and Hall, London, (1956).
(116) M.A. Arbib, *A partial survey of cybernetics in eastern Europe and the Soviet Union*, Behavioral Sci., **11**, 193-216, (1966).
(117) A. Rosenblueth, N. Weiner and J. Bigelow, *Behavior, purpose and teleology*, Phil. Soc. **10** (1), 18-24 (1943).
(118) N. Weiner and A. Rosenblueth, *Conduction of impulses in cardiac muscle*, Arch. Inst. Cardiol. Mex., **16**, 205-265 (1946).
(119) H. von Foerster, editor, *Cybernetics - circular, causal and feed-back mechanisms in biological and social systems. Transactions of sixth-tenth conferences*, Josiah J. Macy Jr. Foundation, New York, (1950-1954).
(120) W.S. McCulloch and W. Pitts, *A logical calculus of ideas immanent in nervous activity*, Bull. Math. Biophys., **5**, 115-133 (1943).
(121) W.S. McCulloch, *An Account of the First Three Conferences on Teleological Mechanisms*, Josiah Macy Jr. Foundation, (1947).
(122) G.A. Miller, *Languages and Communication*, McGraw-Hill, New York, (1951).
(123) G.A. Miller, *Statistical behavioristics and sequences of responses*, Psychol. Rev. **56**, 6 (1949).
(124) G. Bateson, *Bali - the value system of a steady state*, in M. Fortes, editor, *Social Structure Studies Presented to A.R. Radcliffe-Brown*, Clarendon Press, Oxford, (1949).
(125) G. Bateson, *Communication, the Social Matrix of Psychiatry*, Norton, (1951).
(126) G. Bateson, *Steps to an Ecology of Mind*, Chandler, San Francisco, (1972).
(127) G. Bateson, *Communication et Societé*, Seuil, Paris, (1988).
(128) S. Heims, *Gregory Bateson and the mathematicians: From interdisciplinary interactions to societal functions*, J. History Behavioral Sci., **13**, 141-159 (1977).
(129) S. Heims, *John von Neumann and Norbert Wiener. From Mathematics to the Technology of Life and Death*, MIT Press, Cambridge MA, (1980).
(130) S. Heims, *The Cybernetics Group*, MIT Press, Cambridge MA, (1991).
(131) G. van de Vijver, *New Perspectives on Cybernetics (Self-*

Organization, Autonomy and Connectionism), Kluwer, Dordrecht, (1992).
(132) A. Bavelas, *A mathematical model for group structures*, Appl. Anthrop. **7** (3), 16 (1948).
(133) P. de Latil, *La Pensee Artificielle - Introduction a la Cybernetique*, Gallimard, Paris, (1953).
(134) L.K. Frank, G.E. Hutchinson, W.K. Livingston, W.S. McCulloch and N. Wiener, *Teleological Mechanisms*, Ann. N.Y. Acad. Sci. **50**, 187-277 (1948).
(135) H. von Foerster, *Quantum theory of memory*, in H. von Foerster, editor, *Cybernetics - circular, causal and feed-back mechanisms in biological and social systems. Transactions of the sixth conferences*, Josiah J. Macy Jr. Foundation, New York, (1950).
(136) H. von Foerster, *Observing Systems*, Intersystems Publications, California, (1984).
(137) H. von Foerster, *Understanding Understanding: Essays on Cybernetics and Cognition*, Springer, New York, (2002).
(138) M. Newborn, *Kasparov vs. Deep Blue: Computer Chess Comes of age*, Springer Verlag, (1996).
(139) K.M. Colby, *Artificial Paranoia: A Computer Simulation of the Paranoid Process*, Pergamon Press, New York, (1975).
(140) J.Z. Young, *Discrimination and learning in the octopus*, in H. von Foerster, editor, *Cybernetics - circular, causal and feed-back mechanisms in biological and social systems. Transactions of the ninth conference*, Josiah J. Macy Jr. Foundation, New York, (1953).
(141) M.J. Apter and L. Wolpert, *Cybernetics and development. I. Information theory*, J. Theor. Biol. **8**, 244-257 (1965).
(142) H. Atlan, *L'Organization Biologique et la Théorie de l'Information*, Hermann, Paris, (1972).
(143) H. Atlan, *On a formal definition of organization*, J. Theor. Biol. **45**, 295-304 (1974).
(144) H. Atlan, *Organization du vivant, information et auto-organization*, in Volume Symposium 1986 de l'Encylopediea Universalis, pp. 355-361, Paris, (1986).
(145) E.R. Kandel, *Nerve cells and behavior*, Scientific American, **223 nr.1**, 57-70, July, (1970).
(146) E.R. Kandel, *Small systems of neurons*, Scientific American, **241 no.3**, 66-76 (1979).
(147) A.K. Katchalsky et al., *Dynamic patterns of brain cell assemblies*, Neurosciences Res. Prog. Bull., **12 no. 1**, (1974).
(148) G.E. Moore, *Cramming more components onto integrated circuits*, Electronics, April 19, (1965).

(149) P. Gelsinger, P. Gargini, G. Parker and A. Yu, *Microprocessors circa 2000*, IEEE Spectrum, October, (1989).
(150) P. Baron, *On distributed communications networks*, IEEE Trans. Comm. Systems, March (1964).
(151) V.G. Cerf and R.E. Khan, *A protocol for packet network intercommunication*, Trans. Comm. Tech. **COM-22, V 5**, 627-641, May (1974).
(152) L. Kleinrock, *Communication Nets: Stochastic Message Flow and Delay*, McGraw-Hill, New York, (1964).
(153) L. Kleinrock, *Queueing Systems: Vol. II, Computer Applications*, Wiley, New York, /1976).
(154) R. Kahn, editor, *Special Issue on Packet Communication Networks*, Proc. IEEE, **66, No. 11**, November, (1978).
(155) L.G. Roberts, *The evolution of packet switching*, Proc. of the IEEE **66**, 1307-13, (1978).
(156) J. Abbate, *The electrical century: Inventing the web*, Proc. IEEE **87**, November, (1999).
(157) J. Abbate, *Inventing the Internet*, MIT Press, Cambridge MA, (1999).
(158) B. Metcalfe, *Packet Communication*, Peer-to-Peer Communication, San José Calif, (1996).
(159) T. Berners-Lee, *The Original Design and Ultimate Destiny of the World Wide Web by its Inventor*, Harper San Francisco, (1999).
(160) J. Clark, *Netscape Time: The Making of the Billion-Dollar Start-Up That Took On Microsoft*, St. Martin's Press, New York, (1999).
(161) J. Wallace, *Overdrive: Bill Gates and the Race to Control Cyberspace*, Wiley, New York, (1997).
(162) P. Cunningham and F. Froschl, *The Electronic Business Revolution*, Springer Verlag, New York, (1999).
(163) J.L. McKenny, *Waves of Change: Business Evolution Through Information Technology*, Harvard Business School Press, (1995).
(164) M.A. Cosumano, *Competing on Internet Time: Lessons From Netscape and Its Battle with Microsoft*, Free Press, New York, (1998).
(165) F.J. Dyson, *The Sun, the Genome and the Internet: Tools of Scientific Revolutions*, Oxford University Press, (1999).
(166) L. Bruno, *Fiber Optimism: Nortel, Lucent and Cisco are battling to win the high-stakes fiber-optics game*, Red Herring, June (2000).
(167) N. Cochrane, *We're insatiable: Now it's 20 million million bytes a day*, Melbourne Age, January 15, (2001).
(168) K.G. Coffman and A.N. Odlyzko, *The size and growth rate of the Internet*, First Monday, October, (1998).
(169) C.A. Eldering, M.L. Sylla, and J.A. Eisenach, *Is there a Moore's law for bandwidth?*, IEEE Comm. Mag., 2-7, October, (1999).

(170) G. Gilder, *Fiber keeps its promise: Get ready, bandwidth will triple each year for the next 25 years*, Forbes, April 7, (1997).
(171) A.M. Noll, *Does data traffic exceed voice traffic?*, Comm. ACM, 121-124, June, (1999).
(172) B. St. Arnaud, J. Coulter, J. Fitchett, and S. Mokbel, *Architectural and engineering issues for building an optical Internet*, Proc. Soc. Optical Eng. (1998).
(173) M. Weisner, *The computer for the 21st century*, Scientific American, September, (1991).
(174) R. Wright, *Three Scientists and Their Gods*, Time Books, (1988).
(175) S. Nora and A. Minc, *The Computerization of Society*, MIT Press, (1981).
(176) T. Forester, *Computers in the Human Context: Information Theory, Productivity, and People*, MIT Press, (1989).
(177) P. Friedland and L.H. Kedes, *Discovering the secrets of DNA*, Comm. of the ACM, **28**, 1164-1185 (1985).
(178) E.F. Meyer, *The first years of the protein data bank*, Protein Science **6**, 1591-7, July (1997).
(179) C. Kulikowski, *Artificial intelligence in medicine: History, evolution and prospects*, in *Handbook of Biomedical Engineering*, J. Bronzine editor, 181.1-181!18, CRC and IEEE Press, Boca Raton Fla., (2000).
(180) C. Gibas and P. Jambeck, *Developing Bioinformatics Computer Skills*, O'Reily, (2001).
(181) F.L. Carter, *The molecular device computer: point of departure for large-scale cellular automata*, Physica D, **10**, 175-194 (1984).
(182) K.E. Drexler, *Molecular engineering: an approach to the development of general capabilities for molecular manipulation*, Proc. Natl. Acad. Sci USA, **78**, 5275-5278 (1981).
(183) K.E. Drexler, *Engines of Creation*, Anchor Press, Garden City, New York, (1986).
(184) D.M. Eigler and E.K. Schweizer, *Positioning single atoms with a scanning electron microscope*, Nature, **344**, 524-526 (1990).
(185) E.D. Gilbert, editor, *Miniaturization*, Reinhold, New York, (1961).
(186) R.C. Haddon and A.A. Lamola, *The molecular electronic devices and the biochip computer: present status*, Proc. Natl. Acad. Sci. USA, **82**, 1874-1878 (1985).
(187) H.M. Hastings and S. Waner, *Low dissipation computing in biological systems*, BioSystems, **17**, 241-244 (1985).
(188) J.J. Hopfield, J.N. Onuchic and D.N. Beritan, *A molecular shift register based on electron transfer*, Science, **241**, 817-820 (1988).
(189) L. Keszthelyi, *Bacteriorhodopsin*, in *Bioenergetics*, P. P. Gräber and G. Millazo (editors), Birkhäuaer Verlag, Basil Switzerland, (1997).

(190) F.T. Hong, *The bacteriorhodopsin model membrane as a prototype molecular computing element*, BioSystems, **19**, 223-236 (1986).

(191) L.E. Kay, *Life as technology: Representing, intervening and molecularizing*, Rivista di Storia della Scienzia, **II, vol.1**, 85-103 (1993).

(192) A.P. Alivisatos et al., *Organization of 'nanocrystal molecules' using DNA*, Nature, **382**, 609-611, (1996).

(193) T. Bjørnholm et al., *Self-assembly of regioregular, amphiphilic polythiophenes into highly ordered pi-stacked conjugated thin films and nanocircuits*, J. Am. Chem. Soc. **120**, 7643 (1998).

(194) L.J. Fogel, A.J.Owens, and M.J. Walsh, *Artificial Intelligence Through Simulated Evolution*, John Wiley, New York, (1966).

(195) L.J. Fogel, *A retrospective view and outlook on evolutionary algorithms*, in *Computational Intelligence: Theory and Applications, 5th Fuzzy Days*, B. Reusch, editor, Springer-Verlag, Berlin, (1997).

(196) P.J. Angeline, *Multiple interacting programs: A representation for evolving complex behaviors*, Cybernetics and Systems, **29 (8)**, 779-806 (1998).

(197) X. Yao and D.B. Fogel, editors, *Proceedings of the 2000 IEEE Symposium on Combinations of Evolutionary Programming and Neural Networks*, IEEE Press, Piscataway, NJ, (2001).

(198) R.M. Brady, *Optimization strategies gleaned from biological evolution*, Nature **317**, 804-806 (1985).

(199) K. Dejong, *Adaptive system design - a genetic approach*, IEEE Syst. M. **10**, 566-574 (1980).

(200) W.B. Dress, *Darwinian optimization of synthetic neural systems*, IEEE Proc. ICNN **4**, 769-776 (1987).

(201) J.H. Holland, *A mathematical framework for studying learning in classifier systems*, Physica **22 D**, 307-313 (1986).

(202) R.F. Albrecht, C.R. Reeves, and N.C. Steele (editors), *Artificial Neural Nets and Genetic Algotithms*, Springer Verlag, (1993).

(203) L. Davis, editor, *Handbook of Genetic Algorithms*, Van Nostrand Reinhold, New York, (1991).

(204) Z. Michalewicz, *Genetic Algorithms + Data Structures = Evolution Programs*, Springer-Verlag, New York, (1992), second edition, (1994).

(205) K.I. Diamantaris and S.Y. Kung, *Principal Component Neural Networks: Theory and Applications*, John Wiley and Sons, New York.

(206) A. Garliauskas and A. Soliunas, *Learning and recognition of visual patterns by human subjects and artificial intelligence systems*, Informatica, **9 (4)**, (1998).

(207) A. Garliauskas, *Numerical simulation of dynamic synapse-dendrite-soma neuronal processes*, Informatica, **9 (2)**, 141-160, (1998).

(208) U. Seifert and B. Michaelis, *Growing multi-dimensional self-*

organizing maps, International Journal of Knowledge-Based Intelligent Engineering Systems, **2 (1)**, 42-48, (1998).
(209) S. Mitra, S.K. Pal, and M.K. Kundu, *Finger print classification using fuzzy multi-layer perceptron*, Neural Computing and Applications, **2**, 227-233 (1994).
(210) M. Verleysen (editor), *European Symposium on Artificial Neural Networks*, D-Facto, (1999).
(211) R.M. Golden, *Mathematical Methods for Neural Network Analysis and Design*, MIT Press, Cambridge MA, (1996).
(212) S. Haykin, *Neural Networks - (A) Comprehensive Foundation*, MacMillan, New York, (1994).
(213) M.A. Grönroos, *Evolutionary Design of Neural Networks*, Thesis, Computer Science, Department of Mathematical Sciences, University of Turku, Finland, (1998).
(214) D.E. Goldberg, Genetic Algorithms in Search, Optimization and Machine Learning, Addison-Wesley, (1989).
(215) M. Mitchell, *An Introduction to Genetic Algorithms*, MIT Press, Cambridge MA, (1996).
(216) L. Davis (editor), *Handbook of Genetic Algorithms*, Van Nostrand and Reinhold, New York, (1991).
(217) J.H. Holland, *Adaptation in Natural and Artificial Systems*, MIT Press, Cambridge MA, (1992).
(218) J.H. Holland, *Hidden Order; How Adaptation Builds Complexity*, Addison Wesley, (1995).
(219) W. Banzhaf, P. Nordin, R.E. Keller and F. Francone, *Genetic Programming - An Introduction; On the Automatic Evolution of Computer Programs and its Applications*, Morgan Kaufmann, San Francisco CA, (1998).
(220) W. Banzhaf et al. (editors), *(GECCO)-99: Proceedings of the Genetic Evolutionary Computation Conference*, Morgan Kaufman, San Francisco CA, (2000).
(221) W. Banzhaf, *Editorial Introduction*, Genetic Programming and Evolvable Machines, **1**, 5-6, (2000).
(222) W. Banzhaf, *The artificial evolution of computer code*, IEEE Intelligent Systems, **15**, 74-76, (2000).
(223) J.J. Grefenstette (editor), *Proceedings of the Second International Conference on Genetic Algorithms and their Applications*, Lawrence Erlbaum Associates, Hillsdale New Jersey, (1987).
(224) J. Koza, *Genetic Programming: On the Programming of Computers by means of Natural Selection*, MIT Press, Cambridge MA, (1992).
(225) J. Koza et al., editors, *Genetic Programming 1997: Proceedings of the Second Annual Conference*, Morgan Kaufmann, San Francisco, (1997).

(226) W.B. Langdon, *Genetic Programming and Data Structures*, Kluwer, (1998).
(227) D. Lundh, B. Olsson, and A. Narayanan, editors, *Bio-Computing and Emergent Computation 1997*, World Scientific Press, Singapore, (1997).
(228) P. Angeline and K. Kinnear, editors, *Advances in Genetic Programming: Volume 2*, MIT Press, (1997).
(229) J.H. Holland, *Adaptation in Natural and Artificial Systems*, The University of Michigan Press, Ann Arbor, (1975).
(230) David B. Fogel and Wirt Atmar (editors), *Proceedings of the First Annual Conference on Evolutionary Programming*, Evolutionary Programming Society, La Jolla California, (1992).
(231) M. Sipper et al., *A phylogenetic, ontogenetic, and epigenetic view of bioinspired hardware systems*, IEEE Transactions in Evolutionary Computation **1**, 1 (1997).
(232) E. Sanchez and M. Tomassini, editors, *Towards Evolvable Hardware*, Lecture Notes in Computer Science, **1062**, Springer-Verlag, (1996).
(233) J. Markoff, *A Darwinian creation of software*, New York Times, Section C, p.6, February 28, (1990).
(234) A. Thompson, *Hardware Evolution: Automatic design of electronic circuits in reconfigurable hardware by artificial evolution*, Distinguished dissertation series, Springer-Verlag, (1998).
(235) W. McCulloch and W. Pitts, *A Logical Calculus of the Ideas Immanent in Nervous Activity*, Bulletin of Mathematical Biophysics, **7**, 115-133, (1943).
(236) F. Rosenblatt, *Principles of Neurodynamics*, Spartan Books, (1962).
(237) C. von der Malsburg, *Self-Organization of Orientation Sensitive Cells in the Striate Cortex*, Kybernetik, **14**, 85-100, (1973).
(238) S. Grossberg, *Adaptive Pattern Classification and Universal Recoding: 1. Parallel Development and Coding of Neural Feature Detectors*, Biological Cybernetics, **23**, 121-134, (1976).
(239) J.J. Hopfield and D.W. Tank, *Computing with Neural Circuits: A Model*, Science, **233**, 625-633, (1986).
(240) R.D. Beer, *Intelligence as Adaptive Behavior: An Experiment in Computational Neuroethology*, Academic Press, New York, (1990).
(241) S. Haykin, *Neural Networks: A Comprehensive Foundation*, IEEE Press and Macmillan, (1994).
(242) S.V. Kartalopoulos, *Understanding Neural Networks and Fuzzy Logic: Concepts and Applications*, IEEE Press, (1996).
(243) D. Fogel, *Evolutionary Computation: The Fossil Record*, IEEE Press, (1998).
(244) D. Fogel, *Evolutionary Computation: Toward a New Philosophy of Machine Intelligence*, IEEE Press, Piscataway NJ, (1995).

(245) J.M. Zurada, R.J. Marks II, and C.J. Robinson, editors, *Computational Intelligence: Imitating Life*, IEEE Press, (1994).
(246) J. Bezdek and S.K. Pal, editors, *Fuzzy Models for Pattern Recognition: Methods that Search for Structure in Data*, IEEE Press, (1992).
(247) M.M. Gupta and G.K. Knopf, editors, *Neuro-Vision Systems: Principles and Applications*, IEEE Press, (1994).
(248) C. Lau, editor, *Neural Networks. Theoretical Foundations and Analysis*, IEEE Press, (1992).
(249) T. Back, D.B. Fogel and Z. Michalewicz, editors, *Handbook of Evolutionary Computation*, Oxford University Press, (1997).
(250) D.E. Rumelhart and J.L. McClelland, *Parallel Distributed Processing: Explorations in the Microstructure of Cognition, Volumes I and II*, MIT Press, (1986).
(251) J. Hertz, A. Krogh and R.G. Palmer, *Introduction to the Theory of Neural Computation*, Addison Wesley, (1991).
(252) J.A. Anderson and E. Rosenfeld, *Neurocomputing: Foundations of Research*, MIT Press, (1988).
(253) R.C. Eberhart and R.W. Dobbins, *Early neural network development history: The age of Camelot*, IEEE Engineering in Medicine and Biology **9**, 15-18 (1990).
(254) T. Kohonen, *Self-Organization and Associative Memory*, Springer-Verlag, Berlin, (1984).
(255) T. Kohonen, *Self-Organizing Maps*, Springer-Verlag, Berlin, (1997).
(256) G.E. Hinton, *How neural networks learn from experience*, Scientific American **267**, 144-151 (1992).
(257) K. Swingler, *Applying Neural Networks: A Practical Guide*, Academic Press, New York, (1996).
(258) B.K. Wong, T.A. Bodnovich and Y. Selvi, *Bibliography of neural network business applications research: 1988-September 1994*, Expert Systems **12**, 253-262 (1995).
(259) I. Kaastra and M. Boyd, *Designing neural networks for forecasting financial and economic time series*, Neurocomputing **10**, 251-273 (1996).
(260) T. Poddig and H. Rehkugler, *A world model of integrated financial markets using artificial neural networks*, Neurocomputing **10**, 2251-273 (1996).
(261) J.A. Burns and G.M. Whiteside, *Feed forward neural networks in chemistry: Mathematical systems for classification and pattern recognition*, Chem. Rev. **93**, 2583-2601, (1993).
(262) M.L. Action and P.W. Wilding, *The application of backpropagation neural networks to problems in pathology and laboratory medicine*, Arch. Pathol. Lab. Med. **116**, 995-1001 (1992).

(263) D.J. Maddalena, *Applications of artificial neural networks to problems in quantitative structure activity relationships*, Exp. Opin. Ther. Patents **6**, 239-251 (1996).

(264) W.G. Baxt, *Application of artificial neural networks to clinical medicine*, [Review], Lancet **346**, 1135-8 (1995).

(265) A. Chablo, *Potential applications of artificial intelligence in telecommunications*, Technovation **14**, 431-435 (1994).

(266) D. Horwitz and M. El-Sibaie, *Applying neural nets to railway engineering*, AI Expert, 36-41, January (1995).

(267) J. Plummer, *Tighter process control with neural networks*, 49-55, October (1993).

(268) T. Higuchi et al., *Proceedings of the First International Conference on Evolvable Systems: From Biology to Hardware (ICES96)*, Lecture Notes on Computer Science, Springer-Verlag, (1997).

(269) S.A. Kaufman, *Antichaos and adaption*, Scientific American, **265**, 78-84, (1991).

(270) S.A. Kauffman, *The Origins of Order*, Oxford University Press, (1993).

(271) M.M. Waldrop, *Complexity: The Emerging Science at the Edge of Order and Chaos*, Simon and Schuster, New York, (1992).

(272) H.A. Simon, *The Science of the Artificial, 3rd Edition*, MIT Press, (1996).

(273) M.L. Hooper, *Embryonic Stem Cells: Introducing Planned Changes into the Animal Germline*, Harwood Academic Publishers, Philadelphia, (1992).

(274) F. Grosveld, (editor), *Transgenic Animals*, Academic Press, New York, (1992).

(275) G. Köhler and C. Milstein, *Continuous cultures of fused cells secreting antibody of predefined specificity*, Nature, **256**, 495-497, (1975).

(276) S. Spiegelman, *An approach to the experimental analysis of precellular evolution*, Quarterly Reviews of Biophysics, **4**, 213-253, (1971).

(277) M. Eigen, *Self-organization of matter and the evolution of biological macromolecules*, Naturwissenschaften, **58**, 465-523, (1971).

(278) M. Eigen and W. Gardiner, *Evolutionary molecular engineering based on RNA replication*, Pure and Applied Chemistry, **56**, 967-978, (1984).

(279) G.F. Joyce, *Directed molecular evolution*, Scientific American **267** (6), 48-55, (1992).

(280) N. Lehman and G.F. Joyce, *Evolution in vitro of an RNA enzyme with altered metal dependence*, Nature, **361**, 182-185, (1993).

(281) E. Culotta, *Forcing the evolution of an RNA enzyme in the test tube*, Science, **257**, 31 July, (1992).

(282) S.A. Kauffman, *Applied molecular evolution*, Journal of Theoretical Biology, **157**, 1-7, (1992).
(283) H. Fenniri, *Combinatorial Chemistry. A Practical Approach*, Oxford University Press, (2000).
(284) P. Seneci, *Solid-Phase Synthesis and Combinatorial Technologies*, John Wiley & Sons, New York, (2001).
(285) G.B. Fields, J.P. Tam, and G. Barany, *Peptides for the New Millennium*, Kluwer Academic Publishers, (2000).
(286) Y.C. Martin, *Diverse viewpoints on computational aspects of molecular diversity*, Journal of Combinatorial Chemistry, **3**, 231-250, (2001).
(287) C.G. Langton et al., editors, *Artificial Life II: Proceedings of the Workshop on Artificial Life Held in Santa Fe, New Mexico*, Adison-Wesley, Reading MA, (1992).
(288) W. Aspray and A. Burks, eds., *Papers of John von Neumann on Computers and Computer Theory*, MIT Press, (1967).
(289) M. Conrad and H.H. Pattee, *Evolution experiments with an artificial ecosystem*, J. Theoret. Biol., **28**, (1970).
(290) C. Emmeche, *Life as an Abstract Phenomenon: Is Artificial Life Possible?*, in *Toward a Practice of Artificial Systems: Proceedings of the First European Conference on Artificial Life*, MIT Press, Cambridge MA, (1992).
(291) C. Emmeche, *The Garden in the Machine: The Emerging Science of Artificial Life*, Princeton University Press, Princeton NJ, (1994).
(292) S. Levy, *Artificial Life: The Quest for New Creation*, Pantheon, New York, (1992).
(293) K. Lindgren and M.G. Nordahl, *Cooperation and Community Structure in Artificial Ecosystems*, Artificial Life, **1**, 15-38.
(294) P. Husbands and I. Harvey (editors), *Proceedings of the 4th Conference on Artificial Life (ECAL '97)*, MIT Press, (1997)
(295) C.G. Langton, (editor), *Artificial Life: An Overview*, MIT Press, Cambridge MA, (1997).
(296) C.G. Langton, ed., *Artificial Life*, Addison-Wesley, (1987).
(297) A.A. Beaudry and G.F. Joyce, *Directed evolution of an RNA enzyme*, Science, **257**, 635-641, (1992).
(298) D.P. Bartel and J.W. Szostak, *Isolation of new ribozymes from a large pool of random sequences*, Science, **261**, 1411-1418, (1993).
(299) K. Kelly, *Out of Control*, http://ww.kk.org/outofcontrol/index.html, (2002).
(300) K. Kelly, *The Third Culture*, Science, February 13, (1998).
(301) S. Blakeslee, *Computer life-form "mutates" in an evolution experiment, natural selection is found at work in a digital world*, New York Times, November 25, (1997).

(302) M. Ward, *It's life, but not as we know it*, New Scientist, July 4, (1998).
(303) P. Guinnessy, *"Life" crawls out of the digital soup*, New Scientist, April 13, (1996).
(304) L. Hurst and R. Dawkins, *Life in a test tube*, Nature, May 21, (1992).
(305) J. Maynard Smith, *Byte-sized evolution*, Nature, February 27, (1992).
(306) W.D. Hillis, *Intelligence as an Emergent Behavior*, in *Artificial Intelligence*, S. Graubard, ed., MIT Press, (1988).
(307) T.S. Ray, *Evolution and optimization of digital organisms*, in *Scientific Excellence in Supercomputing: The IBM 1990 Contest Prize Papers*, K.R. Billingsly, E. Derohanes, and H. Brown, III, editors, The Baldwin Press, University of Georgia, Athens GA 30602, (1991).
(308) S. Lloyd, *The calculus of intricacy*, The Sciences, October, (1990).
(309) M. Minsky, *The Society of Mind*, Simon and Schuster, (1985).
(310) D. Pines, ed., *Emerging Synthesis in Science*, Addison-Wesley, (1988).
(311) P. Prusinkiewicz and A. Lindenmeyer, *The Algorithmic Beauty of Plants*, Springer-Verlag, (1990).
(312) T. Tommaso and N. Margolus, *Cellular Automata Machines: A New Environment for Modeling*, MIT Press, (1987).
(313) W.M. Mitchell, *Complexity: The Emerging Science at the Edge of Order and Chaos*, Simon and Schuster, (1992).
(314) T.S. Ray, *Artificial Life*, in *From Atoms to Mind*, W. Gilbert amd T.V. Glauco. eds., Istituto della Encyclopedia Italiana Treccani, (Rome) (in press).
(315) T.S. Ray et al., *Kurtzweil's Turing Fallacy*, in *Are We Spiritual Machines?: Ray Kurzweil vs. the Critics of Strong AI*, J. Richards, ed., Viking, (2002).
(316) T.S. Ray, *Aesthetically Evolved Virtual Pets*, in *Artificial Life 7 Workshop Proceedings*, C.C. Maley and E. Bordreau, eds., (2000).
(317) T.S. Ray and J.F. Hart, *Evolution of Differentiation in Digital Organisms*, in *Artificial Life VII, Proceedings of the Seventh International Conference on Artificial Life*, M.A. Bedau, J.S. McCaskill, N.H. Packard, and S. Rasmussen, eds., MIT Press, (2000).
(318) T.S. Ray, *Artificial Life*, in *Frontiers of Life, Vol. 1: The Origins of Life*, R. Dulbecco et al., eds., Academic Press, (2001).
(319) T.S. Ray, *Selecting naturally for differentiation: Preliminary evolutionary results*, Complexity, **3** (5), John Wiley and Sons, (1998).
(320) K. Sims, *Artificial Evolution for Computer Graphics*, Computer Graphics, **25** (4), 319-328, (1991).

Chapter 13

LEARNING TO LIVE IN HARMONY

13.1 New goals for education

Good education ought to make students well adapted to live in their environment. "Environment", in the largest sense, means not only the family setting but also the political, economic and natural environments that surround young people as they grow up today. These environments have changed almost beyond recognition during the last few centuries; in fact, they have changed enormously during the last few decades. The future will bring still more drastic changes. Consequently traditional education is in great need of revision.

When Samuel Johnson visited the Birmingham factory where James Watt's newly-invented steam engines were being manufactured during the first stages of the Industrial Revolution, the owner proudly said to him, "I sell here, Sir, what all the world desires to have - Power!" *Power, Growth, Dominance* and *Profit* have been the traditional ideals of industrial society. However, it is doubtful whether they are appropriate ideals for the future. *Harmony* is a better ideal for the future, and a better goal for education in the world of today.

Adam Smith and other economists of the early Industrial Revolution lived in a world largely empty of human economic activities. They considered the limiting factors in the production of food and goods to be shortages of labor and capital. Natural resources were thought to be present in such large quantities that they were not limiting. In the world-view of the classical economists, growth can continue as long as new capital can be accumulated, and as long as new labor can be supplied by population growth or mechanization. There is no upper limit. However, we are now encountering a new situation.

In recent years the assumptions of the classical economists have become progressively more untrue. It is becoming increasingly obvious that the limiting factors in economic growth are no longer capital, human labor or ingenuity in automation. The limiting factors today are scarce cropland, scarce water, depleted reserves of fossil fuels and mineral resources, vanishing catches from overfished oceans, and limits imposed by the carrying capacity of the environment, by pollution and by climate change.

Future generations will have to adapt their lifestyles to a world of increasing resource scarcity - a world where fossil fuels will become progressively more expensive, finally reaching the point where they will no longer be burned but only used as a starting point for chemical synthesis. The easy life to which the industrialized countries have become accustomed will become progressively more difficult to maintain. New attitudes and new knowledge will be needed. Our educational systems will be faced by the challenge of helping society to adapt to a changed world.

The social impact of science

Science and technology have developed extremely rapidly in recent decades, and they will undoubtedly continue to do so in the future. The result has been that humans now have an unprecedented and constantly increasing power over nature, which can be used for both good and evil. Science has given us the possibility of a life free from hunger and free from the constant threat of death through infectious disease. At the same time, however, our constantly accelerating technology has given us the possibility of destroying civilization in a thermonuclear war.

Since it is almost impossible to prevent science from making new discoveries that can be used both constructively and disastrously, one of the new goals of education must be to give voters the knowledge needed to choose wisely among the ways in which our enormous new powers over nature can to be used. This implies that some familiarity with science is needed even for students who specialize in the humanities. A study of the history and social impact of science ought to be part of the education of both scientists and humanists. This should include discussions of global problems and ethical dilemmas related to scientific and technological progress.

Global ethics

Traditional education has always tried to produce patriotism in its students. This may once have been a reasonable goal, but today a broader view than narrow nationalism is needed. Global interdependence and communication have increased to such an extent that the absolutely sovereign nation-state has become a dangerous anachronism. If the disaster of a third world war is to be avoided, structures of government and law must be built up at an international level. One of the new goals for education should be to prepare students for this great task. Today's students need a global ethic - a loyalty to humanity as a whole, rather that a narrowly nationalistic loyalty.

History has traditionally been taught in such a way that ones own nation is seen as being heroic and always in the right. History textbooks also emphasizes power, dominance and military conflicts. A reformed teaching of history might instead be a chronicle of the gradual cultural advances of humankind as a whole, giving adequate recognition to the contributions of all nations and peoples, and giving weight to constructive achievements rather than to power struggles and conflicts.

13.2 Learning from pre-industrial cultures

The legacy of the Industrial Revolution and colonialism can be seen in the division of today's world into a set of rich industrialized countries (typified by the G8) and a group of much less industrialized countries, with far smaller per capita GDP's. The problem of achieving equal economic conditions throughout the world must be solved by the generation of students now going through our school system, and their education must prepare them for this task. A stable future world cannot be a world of inequality.

The era of colonialism has also left the industrialized countries with a rather arrogant attitude towards other cultures. Although formal political colonialism has almost entirely vanished, many of the assumptions of the colonial era persist and are strongly supported by the mainstream mass media. It is assumed by many people in the industrialized North that if the developing countries would only learn mass production, modern farming techniques and a modern lifestyle, all would be well. However, a sustainable global future may require a transfer of knowledge, techniques and attitudes in precisely the opposite direction - from pre-industrial societies to highly industrialized ones. The reason for this is that the older societies have cultures that allow them to live in harmony with nature, and this is exactly what the highly industrial North must learn to do.

Industrialism and the rapid development of science and technology have given some parts of the world a 200-year period of unbroken expansion and growth, but today this growth is headed for a collision with a wall-like barrier - limits set by the carrying capacity of the global environment and by the exhaustion of non-renewable resources. Encountering these limits is a new experience for the the industrialized countries. By contrast, pre-industrial societies have always experienced limits. The industrialized world must soon replace the economics of growth with equilibrium economics. Pre-industrial societies have already learned to live in equilibrium - in harmony with nature.

Like biodiversity, cultural diversity is an extremely valuable resource, and for similar reasons. A large genetic pool gives living organisms the flexibility needed to adapt to changes in the environment. Similarly, cultural diversity can give humans the flexibility needed to cope with change. In the changed world of today (changed by the invention of thermonuclear weapons and by the extraordinary growth of global population and commerce, soon to be changed still more by the end of the fossil fuel era) we urgently need to learn to live in harmony, in harmony with ourselves, in harmony with nature, and in harmony with other members of our species. We can do this if we draw on the full human heritage of cultural diversity. We can draw not only on the knowledge and wisdom of presently existing societies, but also on the experiences and ideas of societies of the past.

The Pythagorean concept of harmony

In the ancient world, the concept of harmony was developed to a high level by the Pythagoreans. The Pythagoreans used the idea of harmony to understand medicine, art, music, mathematics and ethics.

The concept of harmony in Chinese civilization

Taoist and Confucian teachings each emphasized a particular aspect of harmony. Taoism emphasized harmony with nature, while Confucianism taught harmonious relationships between humans. Thus in China, harmony became an ideal advocated by both traditions.

India

Both Hindu and Buddhist traditions emphasize the unity of all life on earth. Hindus regard killing an animal as a sin, and many try to avoid accidentally stepping on insects as they walk. The Hindu and Buddhist picture of the relatedness of all life on earth has been confirmed by modern biological science. We now know that all living organisms have the same fundamental biochemistry, based on DNA, RNA, proteins and polysaccharides, and we know that our own human genomes are more similar to than different from the genomes of our close relations in the animal world.

The peoples of the industrialized nations urgently need to acquire a non-anthropocentric element in their ethics, similar to reverence for all life found in the Hindu and Buddhist traditions, as well as in the teachings of Saint Francis of Assisi and Albert Schweitzer. We need to learn to value

Fig. 13.1 *Pythagoras (seen here in a painting by Raffaello Sanzio) founded a brotherhood that lasted about a hundred years and greatly influenced the development of philosophy and science. The Pythagoreans searched for harmony in mathematics, medicine, art and human relations.*

other species for their own sakes, and not because we expect to use them for our own economic goals.

The Buddhist concept of karma has great value in human relations. The word "karma" means simply "action". In Buddhism, one believes that actions return to the actor. Good actions will be returned, and bad actions

Fig. 13.2 *This painting illustrates the concept of karma. A lady gives books and clothing to a poor student. Later she receives a gift from a neighbor. There may sometimes be a direct causal connection between such events, but often they are connected only by the fact that each act of kindness makes the world a better place. (Himalayan Academy Publications, Kapaa, Kauai, Hawaii.)*

will also be returned. This is obviously true in social relationships. If we behave with kindness and generosity to our neighbors, they will return our kindness. Conversely, a harmful act may lead to a vicious circle of revenge and counter-revenge which can only be broken by returning good for evil.

Fig. 13.3 *A garden statue showing St. Francis of Assisi (1182-1226). According to legend, St. Francis preached to the birds, whom he regarded as his brothers and sisters. His respect and affection for all life has echos in the religion of the Native Americans.*

However the concept of karma has a broader and more abstract validity beyond the direct return of actions to the actor.

When we perform a good action, we increase the total amount of good karma in the world. If all people similarly behave well, the the world as a whole will become more pleasant and more safe. Human nature seems to have a built-in recognition of this fact, and we are rewarded by inner happiness when we perform good and kind actions. In his wonderful book, "Ancient Wisdom, Modern World", the Dalai Lama says that good actions lead to happiness and bad actions to unhappiness even if our neighbors do not return these actions. Inner peace, he tells us, is incompatible with bad karma and can be achieved only through good karma, i.e. good actions.

Original American societies

Today a few societies still follow a way of life similar to that of our hunter-gatherer ancestors. Anthropologists are able to obtain a vivid picture of the past by studying these societies. Often the religious ethics of the hunter-gatherers emphasizes the importance of harmony with nature. For example, respect for nature appears in the tribal traditions of Native Americans.

The attitude towards nature of the Sioux can be seen from the following quotations from *Land of the Spotted Eagle* by the Lakota (Western Sioux) chief, Standing Bear (ca. 1834 - 1908):

"The Lakota was a true lover of Nature. He loved the earth and all things of the earth... From Waken Tanka (the Great Spirit) there came a great unifying life force that flowered in and through all things – the flowers of the plains, blowing winds, rocks, trees, birds, animals – and was the same force that had been breathed into the first man. Thus all things were kindred and were brought together by the same Great Mystery."

"Kinship with all creatures of the earth, sky, and water was a real and active principle. For the animal and bird world there existed a brotherly feeling that kept the Lakota safe among them. And so close did some of the Lakota come to their feathered and furred friends that in true brotherhood they spoke a common tongue."

"The animal had rights – the right of man's protection, the right to live, the right to multiply, the right to freedom, and the right to man's indebtedness – and in recognition of these rights the Lakota never enslaved the animal, and spared all life that was not needed for food and clothing."

"This concept of life was humanizing and gave to the Lakota an abiding love. It filled his being with the joy and mystery of things; it gave him reverence for all life; it made a place for all things in the scheme of existence with equal importance to all. The Lakota could despise no creature, for all were one blood, made by the same hand, and filled with the essence of the Great Mystery."

A similar attitude towards nature can be found in traditional Inuit cultures.

African respect for nature

In some parts of Africa, a man who plans to cut down a tree offers a prayer of apology, telling the tree why necessity has forced him to harm it. This pre-industrial attitude is something from which the industrialized North could learn. In industrial societies, land "belongs" to some one, and the owner

Fig. 13.4 *Harmony with nature was a part of the Native American religion.*

has the "right" to ruin the land or to kill the communities of creatures living on it if this happens to give some economic advantage, in much the same way that a Roman slaveowner was thought to have the "right" to kill his slaves. Pre-industrial societies have a much less rapacious and much more custodial attitude towards the land and towards its non-human inhabitants.

Traditional agricultural societies

Many traditional agricultural societies have an ethical code that requires them to preserve the fertility of the land for future generations. This recognition of a duty towards the distant future is in strong contrast to the shortsightedness of modern economists.

13.3 Science and social institutions

When he heard of the nuclear destruction of Hiroshima and Nagasaki, Albert Einstein said, "Everything has changed except our way of thinking." New ways of thinking are urgently required to deal with the threat of nuclear weapons. The need for new ways of thinking implies a need for new

Fig. 13.5 *The reader may enjoy the Dalai Lama's splendid book, "Ancient Wisdom, Modern World".*

modes of education. Two enormous tasks for the future will face the students passing through our educational systems. The first of these great tasks will be to stabilize the global population. The second great challenge for the future will be to eliminate the institution of war and to replace it by humane governance and an equitable system of laws at the global level.

If we look several centuries into the future, it becomes clear that the survival of civilization requires that nuclear weapons (and ultimately the institution of war itself) must be abolished.

The explosive growth of science-driven technology during the last two centuries has changed the world completely; and our social and political institutions have adjusted much too slowly to the change. The great problem of our times is to keep society from being shaken to pieces by the headlong progress of science - the problem of harmonizing our social and political institutions with technological change. Because of the great importance of this problem, it is perhaps legitimate to ask whether anyone today can be considered to be educated without having studied the impact of science on society. Should we not include this topic in the education of both scientists and non-scientists?

Science has given us great power over the forces of nature. If wisely used, this power will contribute greatly to human happiness; if wrongly used, it will result in misery. In the words of the Spanish writer, Ortega y Gasset, "We live at a time when man, lord of all things, is not lord of himself"; or as Arthur Koestler has remarked, "We can control the movements of a spaceship orbiting about a distant planet, but we cannot control the situation in Northern Ireland."

Thus, far from being obsolete in a technological age, wisdom and ethics are needed now, more than ever before. We need the ethical insights of the great religions and philosophies of humankind - especially the insight which tells us that all humans belong to a single family, that in fact all living creatures are related, and that even inanimate nature deserves our care and respect.

Modern biology has given us the power to create new species and to exert a drastic influence on the course of evolution; but we must use this power with great caution, and with a profound sense of responsibility. There is a possibility that human activities may cause 20% of all species to become extinct within a few decades if we do not act with restraint. The beautiful and complex living organisms on our planet are the product of more than three billion years of evolution. The delicately balanced and intricately interrelated communities of living things on earth must not be destroyed by human greed and thoughtlessness. We need a sense of evolutionary responsibility - a non-anthropocentric component in our system of ethics.

13.4 Building the future

What kind of world do we want for the future? We want a world where war is abolished as an institution, and where the enormous resources now wasted on war are used constructively. We want a world where a stable population of moderate size lives in comfort and security, free from fear of unemployment and fear of hunger. We want a world where peoples of all countries have equal access to resources, and an equal quality of life. We want a world with a new economic system where the prices of resources are not merely the prices of the burglar's tools needed to crack the safes of nature - a system which is not designed to produce unlimited growth, but which aims instead at meeting the real needs of the human community in equilibrium with the global environment. We want a world of changed values, where extravagance and waste are regarded as morally wrong; where kindness, wisdom and beauty are admired; and where the survival of other species than our own is regarded as an end in itself, not just a means to our own ends. No person can make such changes alone, but together we can build the kind of future that we want.

- *In the world of today, more than a trillion US dollars are spent each year on armaments.*

  In the world of the future, a future that we can build by working together, the enormous sums now wasted on war will be used to combat famine, poverty, illiteracy, and preventable disease.

- *In the world of today, population is increasing so fast that it doubles every thirty-nine years. Most of this increase is in the developing countries, and in many of these, the doubling time is less than twenty-five years. Famine is already present, and it threatens to become more severe and widespread in the future.*

  In the world of the future, a future without fear of famine, population will be stabilized at a level that can be sustained comfortably by the world's food and energy resources. Each country will be responsible for stabilizing its own population, and no country will be allowed to export its problem by sending large numbers of its citizens abroad.

- *In the world of today, the nuclear weapons now stockpiled are sufficient to kill everyone on earth several times over. Nuclear technology is spreading, and many politically unstable countries have recently acquired nuclear weapons or may acquire them soon. Even terrorist groups or organized criminals may acquire such weapons, and there is an increasing danger that they will be used.*

In the world of the future, a war-free future, both the manufacture and the possession of nuclear weapons by individual nations will be prohibited. The same will hold for other weapons of mass destruction.

- *In the world of today, fossil fuels are being used at a rate that implies that they will soon become prohibitively expensive. This is especially true of petroleum and natural gas.*

In the world of the future, a future that will require wholehearted governmental commitment in all nations, research into renewable energy technologies will receive generous funding. Fossil fuel use will be taxed, and renewable energy subsidized.

- *In the world of today, 40% of all research funds are used for projects related to armaments.*

In the world of the future, a future that expresses our longing for peace, research in science and engineering will be redirected towards solving the urgent problems now facing humanity, such as the development of better methods for treating tropical diseases, new energy sources, and new agricultural methods. An expanded UNESCO will replace national military establishments as the patron of science and engineering.

- *In the world of today, gross violations of human rights are common. These include genocide, torture, summary execution, and imprisonment without trial.*

In the world of the future, a future where the arrogance of power is curbed, a system of enforceable international laws will protect individuals against violations of human rights.

- *In the world of today, armaments exported from the industrial countries to the Third World amount to a value of roughly 17 billion dollars per year. This trade in arms increases the seriousness and danger of conflicts in the less developed countries, and diverts scarce funds from their urgent needs.*

  In the world of the future, a future not dominated by cynical profit-seeking at the expense of human suffering, international trade in arms will be strictly limited by enforcible laws.

- *In the world of today, an estimated 10 million children die each year from starvation or from diseases related to malnutrition.*

  In the world of the future, a future where the welfare of children will have a higher place than luxury, the international community will support programs for agricultural development and famine relief on a much larger scale than at present.

- *In the world of today, diarrhea spread by unsafe drinking water kills an estimated 6 million children every year.*

  In the world of the future, a future where human happiness is the goal, the installation of safe and adequate water systems and proper sanitation in all parts of the world will have a high priority and will be supported by ample international funds.

- *In the world of today, malaria, tuberculosis, AIDS, cholera, schistosomiasis, typhoid fever, typhus, trachoma, sleeping sickness and river blindness cause the illness and death of millions of people each year. For example, it is estimated that 200 million people now suffer from schistosomiasis and that 500 million suffer from trachoma, which often causes blindness. In Africa alone, malaria kills more than a million children every year.*

  In the world of the future, a future of cooperation and concern, these preventable diseases will be controlled by a concerted international effort. The World Health Organization will be given sufficient funds to carry out this project.

- *In the world of today, the rate of illiteracy in the 25 least developed countries is 80%. The total number of illiterates in the world is estimated to be 800 million.*

In the world of the future, a future where knowledge and wisdom will be highly respected, the international community will aim at giving all children at least an elementary education. Laws against child labor will prevent parents from regarding very young children as a source of income, thus removing one of the driving forces behind the population explosion. The money invested in education will pay economic dividends after a few years.

- *In the world of today, there is no generally enforcible system of international law, although the International Criminal Court is a step in the right direction.*

In the world of the future, a future governed by law rather than brutal power, the General Assembly of the United Nations will have the power to make international laws. These laws will be binding for all citizens of the world community, and the United Nations will enforce its laws by arresting or fining individual violators, even if they are heads of states. However, the laws of the United Nations will be restricted to international matters, and each nation will run its own internal affairs according to its own laws.

- *In the world of today, each nation considers itself to be "sovereign". In other words, every country considers that it can do whatever it likes, without regard for the welfare of the world community. This means that at the international level we have anarchy.*

In the world of the future, a future of peace and rationality, the concept of national sovereignty will be limited by the needs of the world community. Each nation will decide most issues within its own boundaries, but will yield some of its sovereignty in international matters.

- *In the world of today, the United Nations has no reliable means of raising revenues.*

In the world of the future, a future of global equality and governance, the United Nations will have the power to tax international business transactions, such as exchange of currencies. Each member state will also pay a yearly contribution, and failure to pay will mean loss of voting rights.

- *In the world of today, young men are forced to join national armies, where they are trained to kill their fellow humans. Often, if they refuse for reasons of conscience, they are thrown into prison.*

In the world of the future, a future honoring individual conscience, national armies will be very much reduced in size. A larger force of volunteers will be maintained by the United Nations to enforce international laws. The United Nations will have a monopoly on heavy armaments, and the manufacture or possession of nuclear weapons will be prohibited.

- *In the world of today, young people are indoctrinated with nationalism. History is taught in such a way that one's own nation is seen as heroic and in the right, while other nations are seen as inferior or as enemies.*

In the world of the future, a future of enlightenment and human solidarity, young people will be taught to feel loyalty to humanity as a whole. History will be taught in such a way as to emphasize the contributions that all nations and all races have made to the common cultural heritage of humanity.

- *In the world of today, young people are often faced with the prospect of unemployment. This is true both in the developed countries, where automation and recession produce unemployment, and in the developing countries, where unemployment is produced by overpopulation and by lack of capital.*

In the world of the future, a future of hope, the idealism and energy of youth will be fully utilized by the world community to combat illiteracy and disease, and to develop agriculture and industry in the Third World. These projects will be financed by the UN using revenues derived from taxing international currency transactions.

- *In the world of today, women form more than half of the population, but they are not proportionately represented in positions of political and economic power or in the arts and sciences. In many societies, women are confined to the traditional roles of childbearing and housekeeping.*

In the world of the future, a future of social, legal and economic equality, women in all cultures will take their place beside men in positions of importance in government and industry, and in the arts and sciences. The reduced emphasis on childbearing will help to slow the population explosion.

- *In the world of today, pollutants are dumped into our rivers, oceans and atmosphere. Some progress has been made in controlling pollution, but far from enough.*

In the world of the future, a future of harmony with nature, a stabilized and perhaps reduced population will put less pressure on the environment. Strict international laws will prohibit the dumping of pollutants into our common rivers, oceans and atmosphere. The production of greenhouse gasses will also be limited by international laws.

- *In the world of today, there are no enforcible laws to prevent threatened species from being hunted to extinction. Many indigenous human cultures are also threatened.*

In the world of the future, a future with respect for diversity, an enforcible system of international laws will protect threatened species. Indigenous human cultures will also be protected.

- *In the world of today, large areas of tropical rain forest are being destroyed by excessive timber cutting. The cleared land is generally unsuitable for farming.*

In the world of the future, a future where the human footprint is made as light as possible, it will be recognized that the conversion of carbon dioxide into oxygen by tropical forests is necessary for the earth's climatic stability. Tropical forests will also be highly valued because of their enor-

mous diversity of plant and animal life, and large remaining areas of forest will be protected.

- *In the world of today, terrorists often feel that they can expect protection and help from countries sympathetic with their views.*

  In the world of the future, a future of respect for life, a universal convention against terrorism and hijacking will give terrorists no place to hide.

- *In the world of today, opium poppies and other drug-producing plants are grown with little official hindrance in certain parts of Asia, the Middle East, and Latin America. Hard drugs refined from these plants are imported illegally into the developed countries, where they become a major source of high crime rates and human tragedy.*

  In the world of the future, a future of global cooperation, all nations will work together in a coordinated world-wide program to prevent the growing, refinement and distribution of harmful drugs,

- *In the world of today, modern communications media, such as television, films and newspapers, have an enormous influence on public opinion. However, this influence is only rarely used to build up international understanding and mutual respect.*

  In the world of the future, a future of cross-cultural understanding, mass communications media will be more fully used to bridge human differences. Emphasis will be shifted from the sensational portrayal of conflicts to programs that widen our range of sympathy and understanding.

- *In the world of today, international understanding is blocked by language barriers.*

  In the world of the future, a future of shared knowledge, an international language will be selected, and every child will be taught it as a second language.

- *In the world of today, power and material goods are valued more highly than they deserve to be. "Civilized" life often degenerates into a struggle of all against all for power and possessions. However, the industrial complex on which the production of goods depends cannot be made to run faster and faster, because we will soon encounter shortages of energy and raw materials.*

In the world of the future, a future of changed values, non-material human qualities, such as kindness, politeness, and knowledge, and musical, artistic or literary ability will be valued more highly, and people will derive a larger part of their pleasure from conversation, and from the appreciation of unspoiled nature.

- *In the world of today, the institution of slavery existed for so many millennia that it seemed to be a permanent part of human society. Slavery has now been abolished in almost every part of the world. However war, an even greater evil than slavery, still exists as an established human institution.*

  In the world of the future, a future that will require hard work and dedication, we will take courage from the abolition of slavery, and we will turn with energy and resolution to the great task of abolishing war.

- *In the world of today, people feel anxious about the future, but unable to influence it. They feel that as individuals they have no influence on the large-scale course of events.*

  In the world of the future, a future that can become a reality if we work to make it so, ordinary citizens will realize that collectively they can shape the future. They will join hands and work together for a better world. They will give as much of themselves to peace as peace is worth.

In our reverence for the intricate beauty and majesty of nature, and our respect for the dignity and rights of other humans, we, as builders of the future, can feel united with the great religious and philosophical traditions of mankind, and with the traditional wisdom of our ancestors.

Pictures sent back by the astronauts show the earth as it really is - a small, fragile, beautiful planet, drifting on through the dark immensity of space - our home, where we must learn to live in harmony with nature and with each other.

Suggestions for further reading

(1) R. Christoph, *Pythagoras, His Life, Teaching and Influence*, Cornell University Press, (2005).

(2) W. Burkert, *Lore and Science in Ancient Pythagoreanism*, Harvard University Press, (1972).
(3) Luther Standing Bear, *Land of the Spotted Eagle*, Houghton Mifflin, (1933).
(4) T. Gyatso, HH the Dalai Lama, *Ancient Wisdom, Modern World: Ethics for the New Millennium*, Abacus, London, (1999).
(5) T. Gyatso, HH the Dalai Lama, *How to Expand Love: Widening the Circle of Loving Relationships*, Atria Books, (2005).
(6) J. Rotblat and D. Ikeda, *A Quest for Global Peace*, I.B. Tauris, London, (2007).
(7) M. Gorbachev and D. Ikeda, *Moral Lessons of the Twentieth Century*, I.B. Tauris, London, (2005).
(8) D. Krieger and D. Ikeda, *Choose Hope*, Middleway Press, Santa Monica CA 90401, (2002).
(9) P.F. Knitter and C. Muzaffar, eds., *Subverting Greed: Religious Perspectives on the Global Economy*, Orbis Books, Maryknoll, New York, (2002).
(10) R. Kumar, ed., *Mahatma Gandhi in the Beginning of the 21st Century*, Gyan Publishing House, New Delhi 110002, India, (2006).
(11) M.K. Gandhi, *My Experiment With Truth*, Dover, (1983).
(12) J.A. Kirk. *Martin Luther King, Jr.*, Pearson Longman, London, (2005).
(13) M.L. King, Jr., *Strength to Love*, (1963).
(14) O. Nathan and H. Norton, eds., *Einstein on Peace*, Crown, New York, (1988).
(15) L. Tolstoy, *The Kingdom of God is Within You* (available from Wikisource).
(16) L. Tolstoy, *A Calendar of Wisdom*, translated by Peter Sekirin, Prentace Hall, (1997).
(17) St. Francis of Assisi, *Canticum Fratris Solis* (*Canticle of Brother Sun*).
(18) A. Schweitzer, *Out of My Life and Thought: An Autobiography*, Johns Hopkins University Press, (1933).
(19) A. Schweitzer, *Indian Thought and Its Development*, (1935).
(20) D. Tutu, R.D. Enright and J. North, *Exploring Forgiveness*, (1998).
(21) S. du Boulay, *Tutu: Voice of the Voiceless*, Eerdmans, (1988).
(22) Earth Charter Initiative *The Earth Charter*, www.earthcharter.org
(23) P.B. Corcoran, ed., *The Earth Charter in Action*, KIT Publishers, Amsterdam, (2005).

Appendix A

The Carnot cycle

In this appendix, we will look briefly at a few concepts of thermodynamics that are important for understanding renewable energy. Although not all readers will wish to wade through the equations presented here, the discussion is included for those who have a background and taste for such things.

A.1 Entropy

When the steam engine was invented, various attempts were made to make it more efficient. Scientists also tried to understand theoretically the maximum efficiency of heat engines. Out of these efforts, the science of thermodynamics grew. The pioneers of this field include Carnot, Clausius, Joule, Kelvin, Helmholtz, Maxwell, Boltzmann, Gibbs and Nernst.

Nicolas Léonard Sadi Carnot (1796-1832) was a French engineer who is known for a single work, his book *Reflections on the Motive Power of Fire*, published in 1824. In this book, Carnot invented the concept of entropy, S, a quantity which increases when heat is added to a system The increase in entropy is directly proportional to the amount of heat added and inversely proportional to the absolute temperature. Thus if dQ represents a small amount of heat added to a system at temperature T, then[1]

$$dS = \frac{dQ}{T}$$

Carnot's ideas were later developed and refined by the German theoretical physicist Rudolph Clausius. Clausius imagined a closed system divided into two parts, one at temperature T_1, and the other at a lower temperature, T_2. If a very small amount of heat dQ flows from the hotter part to

[1]The temperature is measured on the absolute (Kelvin) scale. To convert temperatures from Centigrade to Kelvin, one must add 273.15 degrees.

Fig. A.1 *Sadi Carnot (1796-1832).*

the colder part, then the change in entropy of the system will be

$$dS = \frac{dQ}{T_2} - \frac{dQ}{T_1} > 0$$

Clausius realized that the entropy change of the whole system must be positive, since heat always flows from a warmer body to a colder one, and never in the reverse direction. He then proposed a law which has come to be known as the Second Law of Thermodynamics: *In any spontaneous process, the entropy of the universe increases.*

James Clerk Maxwell (1831-1879) and Ludwig Boltzmann (1844-1906) were able to show that heat can be interpreted in terms of the energy of motion of atoms and molecules. Boltzmann was also able to establish a connection between entropy and the statistical probability of a state. If W is the probability of a state, then according to Boltzmann, entropy S is proportional to the logarithm of W:

$$S = k \ln W$$

Here $k = 1.38 \times 10^{-23}$ Joules/Kelvin degree is known as "Boltzmann's constant". Thus the Second Law of Thermodynamics can be interpreted as saying that the universe moves continually from less probable states to more probable ones, which is not at all surprising.

Boltzmann's constant, k, also appears in the law relating the temperature T, the volume V and the pressure P of an ideal gas;

$$PV = NkT$$

In the ideal gas equation, N is the number of gas particles. For real gasses, this relationship holds to a good approximation, but there are small deviations due to the finite size of the gas particles and their interactions with one another.

Entropy can also be interpreted as a measure of disorder. Looked at from this point of view, the Second Law of Thermodynamics states that the disorder of the universe is constantly increasing - a somewhat melancholy conclusion. Of course, living organisms are able to create local order by making use of energy from the sun. This low-entropy solar energy is then degraded into heat by the processes of life, and finally it is transferred by infrared radiation to the cold dust clouds of outer space. In the total process, local order is created on the earth, but the disorder (entropy) of the universe increases.

A.2 The efficiency of heat engines

In his book *Reflections on the Motive Power of Fire*, Carnot also introduced the concept of a high-temperature heat reservoir at constant temperature T_H, and a low-temperature reservoir at constant temperature T_C. The important point about these reservoirs is that they are so large that heat can be added to or withdrawn from them without appreciably changing their temperatures.

Carnot next imagined a cylinder containing a gas. The cylinder has piston at one end, which exerts a certain force to balance the pressure of the gas. Carnot invented a cycle, which starts with the the cylinder containing the gas in contact with the hot reservoir. The cycle has four phases, as shown in the figure:

(1) During the first phase of the cycle, the cylinder remains in contact with the hot reservoir, so that its temperature is held constant at T_H. The force exerted by the piston is very gradually reduced, the gas expands, doing work on the external environment, and meanwhile the gas absorbs heat from the high-temperature reservoir. This phase of the Carnot

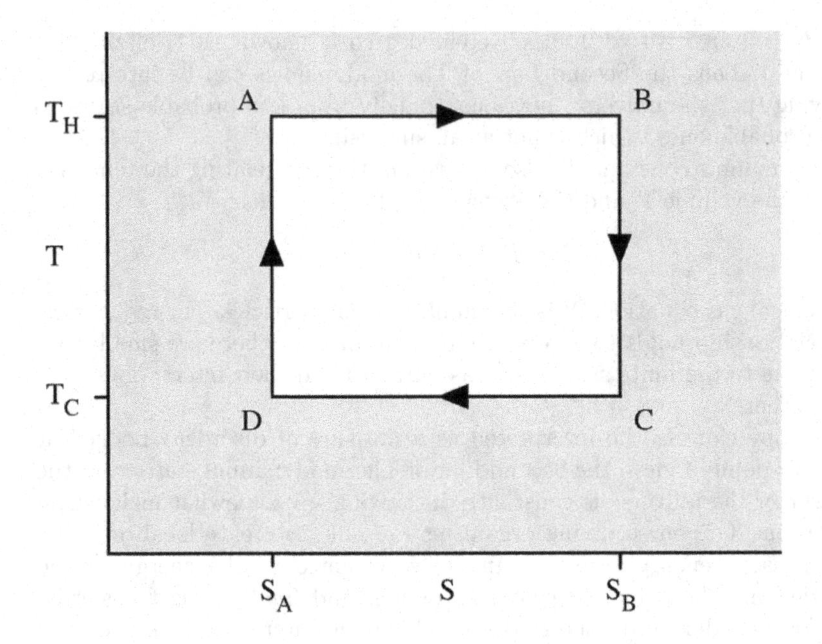

Fig. A.2 *The Carnot cycle.*

cycle is called *Isothermal expansion*, and in it the path is a straight line at temperature T_H, from entropy S_A to entropy S_B. The amount of heat transferred to the gas by the hot reservoir during this phase is given by

$$Q_H = T_H(S_B - S_A)$$

(2) In the second phase of the Carnot cycle, the cylinder is insulated, so that it cannot exchange heat with its surroundings. The force exerted by the piston is gradually reduced still further. The gas expands, doing still more work on the environment. This time the energy required for the external work comes from the cooling of the gas. Finally, the gas becomes so cold that its temperature is T_C. This second phase of the Carnot cycle is called *adiabatic expansion*, and during it, no heat at all is transferred to the gas because it is insulated. Therefore the entropy of the gas is constant during this phase, and the path in the temperature-entropy plans is a straight downward-pointing line from T_H to T_C at constant entropy S_B.

(3) During the third phase of the Carnot cycle, the insulation is removed from the cylinder, and it is placed in contact with the cold reservoir.

Now the force exerted by the piston is gradually increased, and the gas contracts, meanwhile transferring heat to the cold reservoir. Work is being done on the gas by the external world. The third phase of the Carnot cycle is called *isothermal compression*, and in this phase the gas transfers an amount of heat Q_C to the cold reservoir. This amount is given by

$$Q_C = T_C(S_B - S_A)$$

In the temperature-entropy plane, this phase is represented by a horizontal, leftward-pointing line.

(4) In the fourth and final phase of the Carnot cycle, the cylinder is re-insulated, so that no heat can pass in or out of it. Once again, the force exerted by the piston is gradually increased, and the gas is compressed. The compression causes the temperature of the gas to rise, until it finally reaches T_H. The Carnot cycle is complete, and the state of the gas is the same as at the beginning of the cycle. The fourth phase is called *adiabatic compression*. and during it there is no change in entropy. This final phase can be represented by a vertical line from the point (T_C, S_A) to the point (T_H, S_A).

Carnot used his cycle to demonstrate that the maximum efficiency of a heat engine is given by

$$\eta = 1 - \frac{T_C}{T_H}$$

where the temperatures are measured on the Kelvin scale. Carnot noted that the state of a uniform gas is uniquely determined if we know both its temperature and its entropy. Therefore if one makes a graph showing the Carnot cycle as a closed path in a plane with temperature measured along one axis and entropy along the other, the state of the gas is specified at all points along the path.

Thus $Q_H - Q_C$ represents an amount of energy that seems to have disappeared from the system. Where has it gone? According to the First Law of Thermodynamics, energy is conserved, so it must be somewhere. Carnot reasoned that the energy escaped from the system in the form of work performed on the environment by the piston. If this useful work is denoted by the symbol w, then we can write

$$w = Q_H - Q_C$$

The ratio between w and Q_H is the efficiency of the Carnot cycle. It is the ratio of the useful work done to the amount of energy removed from the high-temperature reservoir. If we denote this efficiency by the symbol η, then we have

$$\eta = \frac{w}{Q_H} = \frac{Q_H - Q_C}{Q_H} = 1 - \frac{Q_C}{Q_H} = 1 - \frac{T_C}{T_H}$$

which is Carnot's result. The Carnot cycle is similar to a real heat engine, except that is idealized. The expansions and compressions that occur are reversible - that is to say, the force exerted by the piston exactly balances the pressure of the gas at all stages of the cycle. All real heat engines involve irreversible processes, such as losses to friction, heat losses, etc., and thus they are less efficient than Carnot's idealized heat engine.

When the Carnot cycle is performed in a clockwise direction, it represents a heat engine, from which useful work may be extracted. On the other hand, when the Carnot cycle runs in a counterclockwise direction, it represents a heat pump, which uses work supplied by the external environment to move heat in an unnatural direction, from a colder body to a hotter one. The entropy of the system consisting of the two reservoirs and the gas decreases in this case. This seeming violation of the Second Law of Thermodynamics is only possible because work from the external world is acting on the system through the piston. The entropy of the entire universe still increases. Local order is being created at the expense of disorder elsewhere.

Carnot's expression for maximum efficiency applies both to heat engines and to heat pumps, and it tells us that if we wish to use electricity to heat houses, it is a great mistake to do so by running the electricity through a heat-producing resistor. It is far more efficient to use the electrical power to drive a heat pump, which brings underground heat from a buried network of pipes up into the house. Carnot's efficiency expression tells us that if the temperature difference between the house and the buried pipes is relatively small compared with their absolute (Kelvin) temperatures, very little electrical energy is required to move a large amount of heat.

The Carnot cycle can also help to clarify the difference between fuel energy and electrical energy. If we convert fuels into electrical power by means of heat engines, it must always be at rather low efficiency, as is shown by Carnot's result. Several renewable energy technologies give electrical power directly, and thus they avoid the low efficiency of heat engines. It is much better to use (for example) hydropower of photovoltaics or wind energy to generate electricity than to produce electrical power by burning fuels. On the other hand, if the chemical energy of fuels is converted into electrical power by means of fuel cells, Carnot's efficiency limit does not apply, and the conversion can be much more efficient than it would be if performed by a heat engine.

Appendix B

Dangers of nuclear power generation

B.1 The Chernobyl disaster

The dangers of nuclear power generation are exemplified by the Chernobyl disaster: On the 26th of April, 1986, during the small hours of the morning, the staff of the Chernobyl nuclear reactor in Ukraine turned off several safety systems in order to perform a test. The result was a core meltdown in Reactor 4, causing a chemical explosion that blew off the reactor's 1,000-ton steel and concrete lid. 190 tons of highly radioactive uranium and graphite were hurled into the atmosphere. The resulting radioactive fallout was 200 times greater than that caused by the nuclear bombs that destroyed Hiroshima and Nagasaki. The radioactive cloud spread over Belarus, Ukraine, Russia, Finland, Sweden and Eastern Europe, exposing the populations of these regions to levels of radiation 100 times the normal background. Ultimately, the radioactive cloud reached as far as Greenland and parts of Asia.

The exact number of casualties resulting from the Chernobyl meltdown is a matter of controversy, but according to a United Nations report, as many as 9 million people have been adversely affected by the disaster. Since 1986, the rate of thyroid cancer in affected areas has increased tenfold. An area of 155,000 square kilometers (almost half the size of Italy) in Belarus, Ukraine and Russia is still severely contaminated. Even as far away as Wales, hundreds of farms are still under restrictions because of sheep eating radioactive grass.

Public opinion turned against nuclear power generation as a result of the Chernobyl disaster. Had the disaster taken place in Western Europe or North America, its effect on public opinion would have been still greater. Nevertheless, because of the current energy crisis, and because of worries about global warming, a number of people are arguing that nuclear energy should be given a second chance. The counter-argument is that a

Fig. B.1 *The Chernobyl disaster - much land was made permanently unusable.*

large increase in the share of nuclear power in the total spectrum of energy production would have little effect on climate change but it would involve unacceptable dangers, not only dangers of accidents and dangers associated with radioactive waste disposal, but above all, dangers of proliferation of nuclear weapons.

Of the two bombs that destroyed Hiroshima and Nagasaki, one made use of the rare isotope of uranium, U-235, while the other used plutonium. Both of these materials can be made by a nation with a nuclear power generation program.

Fig. B.2 *Hiroshima*

B.2 Reactors and nuclear weapons

Uranium has atomic number 92, i.e., a neutral uranium atom has a nucleus containing 92 positively-charged protons, around which 92 negatively-charged electrons circle. All of the isotopes of uranium have the same number of protons and electrons, and hence the same chemical properties, but they differ in the number of neutrons in their nuclei. For example, the nucleus of U-235 has 143 neutrons, while that of U-238 has 146. Notice that 92+143=235, while 92+146=238. The number written after the name of an element to specify a particular isotope is the number of neutrons plus the number of protons. This is called the "nucleon number", and the weight of an isotope is roughly proportional to it. This means that U-238 is slightly heavier than U-235. If the two isotopes are to be separated, difficult physical methods dependent on mass must be used, since their chemical properties are identical. In natural uranium, the amount of the rare isotope U-235 is only 0.7 %.

A paper published in 1939 by Niels Bohr and John A. Wheeler indicated that it was the rare isotope of uranium, U-235, that undergoes fission. A bomb could be constructed, they pointed out, if enough highly enriched U-235 could be isolated from the more common isotope, U-238 Calculations later performed in England by Otto Frisch and Rudolf Peierls showed that the "critical mass" of highly enriched uranium needed is quite small - only a few kilograms.

The Bohr-Wheeler theory also predicted that an isotope of plutonium,

Fig. B.3 *Hiroshima*

Pu-239, should be just as fissionable as U-235[1]. Instead of trying to separate the rare isotope, U-235, from the common isotope, U-238, physicists could just operate a nuclear reactor until a sufficient amount of Pu-239 accumulated, and then separate it out by ordinary chemical means.

Thus in 1942, when Enrico Fermi and his coworkers at the University of

[1]Both U-235 and Pu-239 have odd nucleon numbers. When U-235 absorbs a neutron, it becomes U-236, while when Pu-239 absorbs a neutron it becomes Pu-240. In other words, absorption of a neutron converts both these species to nuclei with even nucleon numbers. According to the Bohr-Wheeler theory, nuclei with even nucleon numbers are especially tightly-bound. Thus absorption of a neutron converts U-235 to a highly-excited state of U-236, while Pu-239 is similarly converted to a highly excited state of Pu-240. The excitation energy distorts the nuclei to such an extent that fission becomes possible.

Chicago produced the world's first controlled chain reaction within a pile of cans containing ordinary (nonenriched) uranium powder, separated by blocks of very pure graphite, the chain-reacting pile had a double significance: It represented a new source of energy for mankind, but it also had a sinister meaning. It represented an easy path to nuclear weapons, since one of the by-products of the reaction was a fissionable isotope of plutonium, Pu-239. The bomb dropped on Hiroshima in 1945 used U-235, while the Nagasaki bomb used Pu-239.

By reprocessing spent nuclear fuel rods, using ordinary chemical means, a nation with a power reactor can obtain weapons-usable Pu-239. Even when such reprocessing is performed under international control, the uncertainty as to the amount of Pu-239 obtained is large enough so that the operation might superficially seem to conform to regulations while still supplying enough Pu-239 to make many bombs.

The enrichment of uranium[2] is also linked to reactor use. Many reactors of modern design make use of low enriched uranium (LEU) as a fuel. Nations operating such a reactor may claim that they need a program for uranium enrichment in order to produce LEU for fuel rods. However, by operating their ultracentrifuges a little longer, they can easily produce highly enriched uranium (HEU), i.e., uranium containing a high percentage of the rare isotope U-235, and therefore usable in weapons.

Known reserves of uranium are only sufficient for the generation of 8×10^{20} joules of electrical energy [3], i.e., about 25 TWy. It is sometimes argued that a larger amount of electricity could be obtained from the same amount of uranium through the use of fast breeder reactors, but this would involve totally unacceptable proliferation risks. In fast breeder reactors, the fuel rods consist of highly enriched uranium. Around the core, is an envelope of natural uranium. The flux of fast neutrons from the core is sufficient to convert a part of the U-238 in the envelope into Pu-239, a fissionable isotope of plutonium.

Fast breeder reactors are prohibitively dangerous from the standpoint of nuclear proliferation because both the highly enriched uranium from the fuel rods and the Pu-239 from the envelope are directly weapons-usable. It would be impossible, from the standpoint of equity, to maintain that some nations have the right to use fast breeder reactors, while others do not. If all nations used fast breeder reactors, the number of nuclear weapons states would increase drastically.

It is interesting to review the way in which Israel, South Africa, Pak-

[2]i.e. production of uranium with a higher percentage of U-235 than is found in natural uranium.

[3]Craig, J.R., Vaugn, D.J. and Skinner, B.J., *Resources of the Earth: Origin, Use and Environmental Impact, Third Edition*, page 210.

istan, India and North Korea[4] obtained their nuclear weapons, since in all these cases the weapons were constructed under the guise of "atoms for peace", a phrase that future generations may someday regard as being tragically self-contradictory.

Israel began producing nuclear weapons in the late 1960's (with the help of a "peaceful" nuclear reactor provided by France, and with the tacit approval of the United States) and the country is now believed to possess 100-150 of them, including neutron bombs. Israel's policy is one of visibly possessing nuclear weapons while denying their existence.

South Africa, with the help of Israel and France, also weaponized its civil nuclear program, and it tested nuclear weapons in the Indian Ocean in 1979. In 1991 however, South Africa destroyed its nuclear weapons and signed the NPT.

India produced what it described as a "peaceful nuclear explosion" in 1974. By 1989 Indian scientists were making efforts to purify the lithium-6 isotope, a key component of the much more powerful thermonuclear bombs. In 1998, India conducted underground tests of nuclear weapons, and is now believed to have roughly 60 warheads, constructed from Pu-239 produced in "peaceful" reactors.

Pakistan's efforts to obtain nuclear weapons were spurred by India's 1974 "peaceful nuclear explosion". As early as 1970, the laboratory of Dr. Abdul Qadeer Khan, (a metallurgist who was to become Pakistan's leading nuclear bomb maker) had been able to obtain from a Dutch firm the high-speed ultracentrifuges needed for uranium enrichment. With unlimited financial support and freedom from auditing requirements, Dr. Khan purchased restricted items needed for nuclear weapon construction from companies in Europe and the United States. In the process, Dr. Khan became an extremely wealthy man. With additional help from China, Pakistan was ready to test five nuclear weapons in 1998. The Indian and Pakistani nuclear bomb tests, conducted in rapid succession, presented the world with the danger that these devastating weapons would be used in the conflict over Kashmir. Indeed, Pakistan announced that if a war broke out using conventional weapons, Pakistan's nuclear weapons would be used "at an early stage".

In Pakistan, Dr. A.Q. Khan became a great national hero. He was presented as the person who had saved Pakistan from attack by India by creating Pakistan's own nuclear weapons. In a Washington Post article[5] Pervez Hoodbhoy wrote: "Nuclear nationalism was the order of the day as governments vigorously promoted the bomb as the symbol of Pakistan's

[4]Israel, India and Pakistan have refused to sign the Nuclear Non-Proliferation Treaty, and North Korea, after signing the NPT, withdrew from it in 2003.

[5]1 February, 2004.

high scientific achievement and self-respect..." Similar manifestations of nuclear nationalism could also be seen in India after India's 1998 bomb tests.

Early in 2004, it was revealed that Dr. Khan had for years been selling nuclear secrets and equipment to Libya, Iran and North Korea, and that he had contacts with Al Qaeda. However, observers considered that it was unlikely that Khan would be tried, since a trial might implicate Pakistan's army as well as two of its former prime ministers.

Recent assassination attempts directed at Pakistan's President, Pervez Musharraf, emphasize the precariousness of Pakistan's government. There a danger that it may be overthrown, and that the revolutionists would give Pakistan's nuclear weapons to a subnational organization. This type of danger is a general one associated with nuclear proliferation. As more and more countries obtain nuclear weapons, it becomes increasingly likely that one of them will undergo a revolution, during the course of which nuclear weapons will fall into the hands of criminals or terrorists.

If nuclear reactors become the standard means for electricity generation as the result of a future energy crisis, the number of nations possessing nuclear weapons might ultimately be as high as 40. If this should happen, then over a long period of time the chance that one or another of these nations would undergo a revolution during which the weapons would fall into the hands of a subnational group would gradually grow into a certainty.

There is also a possibility that poorly-guarded fissionable material could fall into the hands of subnational groups, who would then succeed in constructing their own nuclear weapons. Given a critical mass of highly-enriched uranium, a terrorist group, or an organized criminal (Mafia) group, could easily construct a crude gun-type nuclear explosive device. Pu-239 is more difficult to use since it is highly radioactive, but the physicist Frank Barnaby believes that a subnational group could nevertheless construct a crude nuclear bomb (of the Nagasaki type) from this material.

We must remember the remark of U.N. Secretary General Kofi Annan after the 9/11/2001 attacks on the World Trade Center. He said, "*This time* it was not a nuclear explosion". The meaning of his remark is clear: If the world does not take strong steps to eliminate fissionable materials and nuclear weapons, it will only be a matter of time before they will be used in terrorist attacks on major cities, or by organized criminals for the purpose of extortion. Neither terrorists nor organized criminals can be deterred by the threat of nuclear retaliation, since they have no territory against which such retaliation could be directed. They blend invisibly into the general population. Nor can a "missile defense system" prevent criminals or terrorists from using nuclear weapons, since the weapons can be brought into a port in any one of the hundreds of thousands of containers that enter on ships each year, a number far too large to be checked exhaustively.

Finally we must remember that if the number of nations possessing nuclear weapons becomes very large, there will be a greatly increased chance that these weapons will be used in conflicts between nations, either by accident or through irresponsible political decisions.

On November 3, 2003, Mohamed ElBaradei, Director General of the International Atomic Energy Agency, made a speech to the United Nations in which he called for "limiting the processing of weapons-usable material (separated plutonium and high enriched uranium) in civilian nuclear programs - as well as the production of new material through reprocessing and enrichment - by agreeing to restrict these operations to facilities exclusively under international control." It is almost incredible, considering the dangers of nuclear proliferation and nuclear terrorism, that such restrictions were not imposed long ago.

From the facts that we have been reviewing, we can conclude that if nuclear power generation becomes widespread during a future energy crisis, and if equally widespread proliferation of nuclear weapons is to be avoided, the powers and budget of the IAEA will have to be greatly increased. All enrichment of uranium and reprocessing of fuel rods throughout the world will have to be placed be under direct international control, as has been emphasized by Mohamed ElBaradei. Because this will need to be done with fairness, such regulations will have to hold both in countries that at present have nuclear weapons and in countries that do not. It has been proposed that there should be an international fuel rod bank, to supply new fuel rods and reprocess spent ones. In addition to this excellent proposal, one might also consider a system where all power generation reactors and all research reactors would be staffed by the IAEA.

Nuclear reactors used for "peaceful" purposes unfortunately also generate fissionable isotopes of not only of plutonium, but also of neptunium and americium. Thus all nuclear reactors must be regarded as ambiguous in function, and all must be put under strict international control. One must ask whether globally widespread use of nuclear energy is worth the danger that it entails.

Let us now examine the question of whether nuclear power generation would appreciably help to prevent global warming. The fraction of nuclear power in the present energy generation spectrum is at present approximately 1/16. Nuclear energy is used primarily for electricity generation. Thus increasing the nuclear fraction would not affect the consumption of fossil fuels used directly in industry, transportation, in commerce, and in the residential sector. Coal is still a very inexpensive fuel, and an increase in nuclear power generation would do little to prevent it from being burned. Thus besides being prohibitively dangerous, and besides being unsustainable in the long run (because of finite stocks of uranium and thorium), the large-scale use of nuclear power cannot be considered to be a solution to the problem of anthropogenic climate change.

Suggestions for further reading

(1) J.L. Henderson, *Hiroshima*, Longmans (1974).
(2) A. Osada, *Children of the A-Bomb, The Testament of Boys and Girls of Hiroshima*, Putnam, New York (1963).
(3) M. Hachiya, M.D., *Hiroshima Diary*, The University of North Carolina Press, Chapel Hill, N.C. (1955).
(4) M. Yass, *Hiroshima*, G.P. Putnam's Sons, New York (1972).
(5) R. Jungk, *Children of the Ashes*, Harcourt, Brace and World (1961).
(6) B. Hirschfield, *A Cloud Over Hiroshima*, Baily Brothers and Swinfin Ltd. (1974).
(7) J. Hersey, *Hiroshima*, Penguin Books Ltd. (1975).
(8) R. Rhodes, *Dark Sun: The Making of the Hydrogen Bomb*, Simon and Schuster, New York, (1995)
(9) R. Rhodes, *The Making of the Atomic Bomb*, Simon and Schuster, New York, (1988).
(10) D.V. Babst et al., *Accidental Nuclear War: The Growing Peril*, Peace Research Institute, Dundas, Ontario, (1984).
(11) S. Britten, *The Invisible Event: An Assessment of the Risk of Accidental or Unauthorized Detonation of Nuclear Weapons and of War by Miscalculation*, Menard Press, London, (1983).
(12) M. Dando and P. Rogers, *The Death of Deterrence*, CND Publications, London, (1984).
(13) N.F. Dixon, *On the Psychology of Military Incompetence*, Futura, London, (1976).
(14) D. Frei and C. Catrina, *Risks of Unintentional Nuclear War*, United Nations, Geneva, (1982).
(15) H. L'Etang, *Fit to Lead?*, Heinemann Medical, London, (1980).
(16) SPANW, *Nuclear War by Mistake - Inevitable or Preventable?*, Swedish Physicians Against Nuclear War, Lulea, (1985).
(17) J. Goldblat, *Nuclear Non-proliferation: The Why and the Wherefore*, (SIPRI Publications), Taylor and Francis, (1985).
(18) IAEA, *International Safeguards and the Non-proliferation of Nuclear Weapons*, International Atomic Energy Agency, Vienna, (1985).
(19) J. Schear, ed., *Nuclear Weapons Proliferation and Nuclear Risk*, Gower, London, (1984).
(20) D.P. Barash and J.E. Lipton, *Stop Nuclear War! A Handbook*, Grove Press, New York, (1982).

(21) C.F. Barnaby and G.P. Thomas, eds., *The Nuclear Arms Race: Control or Catastrophe*, Francis Pinter, London, (1982).
(22) L.R. Beres, *Apocalypse: Nuclear Catastrophe in World Politics*, Chicago University press, Chicago, IL, (1980).
(23) F. Blackaby et al., eds., *No-first-use*, Taylor and Francis, London, (1984).
(24) NS, ed., *New Statesman Papers on Destruction and Disarmament* (NS Report No. 3), New Statesman, London, (1981).
(25) H. Caldicot, *Missile Envy: The Arms Race and Nuclear War*, William Morrow, New York, (1984).
(26) R. Ehrlich, *Waging the Peace: The Technology and Politics of Nuclear Weapons*, State University of New York Press, Albany, NY, (1985).
(27) W. Epstein, *The Prevention of Nuclear War: A United Nations Perspective*, Gunn and Hain, Cambridge, MA, (1984).
(28) W. Epstein and T. Toyoda, eds., *A New Design for Nuclear Disarmament*, Spokesman, Nottingham, (1975).
(29) G.F. Kennan, *The Nuclear Delusion*, Pantheon, New York, (1983).
(30) R.J. Lifton and R. Falk, *Indefensible Weapons: The Political and Psychological Case Against Nuclearism*, Basic Books, New York, (1982).
(31) J.R. Macy, *Despair and Personal Power in the Nuclear Age*, New Society Publishers, Philadelphia, PA, (1983).
(32) A.S. Miller et al., eds., *Nuclear Weapons and Law*, Greenwood Press, Westport, CT, (1984).
(33) MIT Coalition on Disarmament, eds., *The Nuclear Almanac: Confronting the Atom in War and Peace*, Addison-Wesley, Reading, MA, (1984).
(34) UN, *Nuclear Weapons: Report of the Secretary-General of the United Nations*, United Nations, New York, (1980).
(35) IC, *Proceedings of the Conference on Understanding Nuclear War*, Imperial College, London, (1980).
(36) B. Russell, *Common Sense and Nuclear Warfare*, Allen and Unwin, London, (1959).
(37) F. Barnaby, *The Nuclear Age*, Almqvist and Wiksell, Stockholm, (1974).
(38) D. Albright, F. Berkhout and W. Walker, *Plutonium and Highly Enriched Uranium 1996: World Inventories, Capabilities and Policies*, Oxford University Press, Oxford, (1997).
(39) G.T. Allison et al., *Avoiding Nuclear Anarchy: Containing the Threat of Loose Russian Nuclear Weapons and Fissile Material*, MIT Press, Cambridge MA, (1996).
(40) B. Bailin, *The Making of the Indian Atomic Bomb: Science, Secrecy, and the Post-colonial State*, Zed Books, London, (1998).

(41) G.K. Bertsch and S.R. Grillot, (Eds.), *Arms on the Market: Reducing the Risks of Proliferation in the Former Soviet Union*, Routledge, New York, (1998).
(42) P. Bidawi and A. Vanaik, *South Asia on a Short Fuse: Nuclear Politics and the Future of Global Disarmament*, Oxford University Press, Oxford, (2001).
(43) F.A. Boyle, *The Criminality of Nuclear Deterrence: Could the U.S. War on Terrorism Go Nuclear?*, Clarity Press, Atlanta GA, (2002).
(44) G. Burns, *The Atomic Papers: A Citizen's Guide to Selected Books and Articles on the Bomb, the Arms Race, Nuclear Power, the Peace Movement, and Related Issues*, Scarecrow Press, Metuchen NJ, (1984).
(45) L. Butler, *A Voice of Reason*, The Bulletin of Atomic Scientists, **54**, 58-61, (1998).
(46) R. Butler, *Fatal Choice: Nuclear Weapons and the Illusion of Missile Defense*, Westview Press, Boulder CO, (2001).
(47) R.P. Carlisle (Ed.), *Encyclopedia of the Atomic Age*, Facts on File, New York, (2001).
(48) G.A. Cheney, *Nuclear Proliferation: The Problems and Possibilities*, Franklin Watts, New York, (1999).
(49) A. Cohen, *Israel and the Bomb*, Colombia University Press, New York, (1998).
(50) S.J. Diehl and J.C. Moltz, *Nuclear Weapons and Nonproliferation: A Reference Handbook*, ABC-Clio Information Services, Santa Barbara CA, (2002).
(51) H.A. Feiveson (Ed.), *The Nuclear Turning Point: A Blueprint for Deep Cuts and De-Alerting of Nuclear Weapons*, Brookings Institution Press, Washington D.C., (1999).
(52) R. Forsberg et al., *Nonproliferation Primer: Preventing the Spread of Nuclear, Chemical and Biological Weapons*, MIT Press, Cambridge, (1995).
(53) R. Hilsman, *From Nuclear Military Strategy to a World Without War: A History and a Proposal*, Praeger Publishers, Westport, (1999).
(54) International Physicians for the Prevention of Nuclear War and The Institute for Energy and Environmental Research *Plutonium: Deadly Gold of the Nuclear Age*, International Physicians Press, Cambridge MA, (1992).
(55) R.W. Jones and M.G. McDonough, *Tracking Nuclear Proliferation: A Guide in Maps and Charts, 1998*, The Carnegie Endowment for International Peace, Washington D.C., (1998).
(56) R.J. Lifton and R. Falk, *Indefensible Weapons: The Political and Psychological Case Against Nuclearism*, Basic Books, New York, (1982).

(57) R.E. Powaski, *March to Armageddon: The United States and the Nuclear Arms Race, 1939 to the Present*, Oxford University Press, (1987).

(58) J. Rotblat, J. Steinberger and B. Udgaonkar (Eds.), *A Nuclear-Weapon-Free World: Desirable? Feasible?*, Westview Press, (1993).

(59) The United Methodist Council of Bishops, *In Defense of Creation: The Nuclear Crisis and a Just Peace*, Graded Press, Nashville, (1986).

(60) U.S. Congress Office of Technology Assessment (Ed.), *Dismantling the Bomb and Managing the Nuclear Materials*, U.S. Government Printing Office, Washington D.C., (1993).

(61) S.R. Weart, *Nuclear Fear: A History of Images*, Harvard University Press, (1988).

(62) P. Boyer, *By the Bomb's Early Light: American Thought and Culture at the Dawn of the Atomic Age*, University of North Carolina Press, (1985).

(63) A. Makhijani and S. Saleska, *The Nuclear Power Deception: Nuclear Mythology From Electricity 'Too Cheap to Meter' to 'Inherently Safe' Reactors*, Apex Press, (1999).

(64) C. Perrow, *Normal Accidents: Living With High-Risk Technologies*, Basic Books, (1984).

(65) P. Rogers, *The Risk of Nuclear Terrorism in Britain*, Oxford Research Group, Oxford, (2006).

(66) MIT, *The Future of Nuclear Power: An Interdisciplinary MIT Study*, http://web.mit.edu/nuclearpower, (2003).

(67) Z. Mian and A. Glaser, *Life in a Nuclear Powered Crowd*, INES Newsletter No. 52, 9-13, April, (2006).

(68) K. Bergeron, *Nuclear Weapons: The Death of No Dual-use*, Bulletin of the Atomic Scientists, 15-17, January/February, (2004).

(69) C. Perrow, *Normal Accidents: Living With High-Risk Technologies*, Basic Books, (1984).

(70) K. Bergeron, *Nuclear Weapons: The Death of No-Dual-Use*, Bulletin of Atomic Scientists, January/February, (2004).

(71) P. Rogers, *The Risk of Nuclear Terrorism in Britain*, Oxford Research Group, Oxford, (2006).

Index